UNITED STATES FIRE ADMINISTRATION
FEDERAL EMERGENCY MANAGEMENT AGENCY

THE RURAL FIRE PROBLEM IN THE UNITED STATES

This publication was produced under Contract EMW-94-C-4443 for the United States Fire Administration, Federal Emergency Management Agency. Any information, findings, conclusions, or recommendations expressed in this publication do not necessarily reflect the views of the Federal Emergency Management Agency or the United States Fire Administration.

AUGUST 1997

TABLE OF CONTENTS

EXECUTIVE SUMMARY

This report summarizes the findings from an extensive analysis of the fire problem in rural areas of the U.S. While there are many similarities between fires in rural and non-rural areas, there are also many differences. Some of the differences, such as the higher incidence of heating fires in rural areas, point to issues that need to be considered when designing public education programs to reduce the number of fires and the deaths, injuries, and property loss associated with rural fires. For the purposes of this report, "rural" is defined as all counties that have populations of fewer than 20,000 persons and that are generally not adjacent to metropolitan areas.

The most important findings of this study are summarized below:

Rural Fires

- The leading cause of fires is different in rural areas than in non-rural areas. *Heating* is the leading cause of residential structure fires in rural areas and causes 34 percent of rural residential fires. Heating is the cause of only 15 percent of residential fires in non-rural areas. In contrast, cooking is the leading cause of residential fires in non-rural areas.

- *In particular, the lack of maintenance of heating devices is a serious cause of residential heating fires in rural areas.* Lack of maintenance includes creosote build-up in chimneys and stovepipes. Lack of maintenance was cited in 78 percent of rural heating fires. This suggests a critical need for public education in rural areas to make people aware of the hazards of not properly maintaining heating equipment, chimneys, and vents.

- Stationary heating units are the leading type of equipment involved in ignition of rural residential heating fires. Chimneys, vents, and flues are the second leading category of equipment involved in ignition. Together, these two types of equipment account for 62 percent of all rural heating fires. *Interestingly, "fixed stationary" rather than "portable" heaters are identified as the culprit in this analysis.*

- Because of the prevalence of heating fires, the most common area of fire origin in rural fires is *chimneys*. The next most common areas are cooking areas and lounge areas. Heating equipment rooms are identified as the area of fire origin in only a small proportion of rural heating fires, suggesting that most rural heating fires are not related to central heating.

- Heating is the leading cause of residential fires in rural areas of both the northern and the southern states. Because of the climate, however, heating is a more predominant cause in the North.

- *The lack of working smoke detectors is a significant problem in rural areas.* Smoke detectors were present and operational in only 27 percent of rural residential fires (versus 35 percent of non-rural fires).

- The lack of working smoke detectors is an even greater problem in rural areas of the South than in rural areas of the North.

- The extent of flame damage that residential structures sustain is worse in rural areas than in non-rural areas. This is likely due to two factors. Emergency response times are longer in rural areas due to longer travel distances. Additionally, fires may burn longer before being noticed in rural areas due to lower population densities.

- The leading causes of fires in manufactured housing in rural areas are similar to other types of rural residences.

- The lack of working smoke detectors in manufactured housing is a significant problem. Seventy-five percent of rural manufactured homes that experienced fires do not have an operating smoke detector.

Rural Fire Deaths

- Fire death rates are significantly higher (35 percent higher) in rural areas compared to non-rural areas. These differences are even greater when comparing fire death rates across race and ethnicity groups.

- Within rural areas, the majority of annual fire death victims are White. In per capita terms, however, African Americans and Native Americans have higher risks of dying as a result of fires than do Whites.

- While the death rate is higher in rural areas and for certain subgroups of the population, the distributions of fire deaths by age, race, and gender are similar in "rural" and "non-rural" areas.

INTRODUCTION

This report documents the nature of the fire problem in rural areas of the United States. While other studies have explored various aspects of rural fires, this report provides an analysis of both rural fire deaths and the unique characteristics of fires that occur in rural areas. For the purposes of this report, "rural" is defined as all counties that have populations of fewer than 20,000 persons and that are generally not adjacent to metropolitan areas.

The report is divided into two major parts. The first part uses data from the U.S. Fire Administration's National Fire Incident Reporting System (NFIRS) to delineate the character of rural fires. The second part explores rural fire deaths and how they are distributed by age, race, and gender

Part I of this report is organized into five sections. The first section provides an overview of where rural fires occur. The second section describes the characteristics of the rural fire problem. As the majority of fire casualties occur in residential structures, the third section of Part I deals exclusively with rural residential structures. Section four discusses differences in rural fires between northern and southern areas of the U.S. Finally, the last section of Part I discusses the rural fire problem as it relates to a specific type of housing, manufactured housing.

Part II of this report analyzes fire deaths that result from rural fires. This section addresses both the number of fire deaths and per capita fire death rates by race and ethnicity in rural versus non-rural areas. The data for this analysis are mortality data from the National Center for Health Statistics. The results show that per capita fire death rates are significantly higher in rural than in non-rural areas. More Whites die in fires in rural areas than members of any other race or ethnic group. However, on a per capita basis, the rural residents most at risk of dying in fires are African Americans and Native Americans.

Several appendices provide further data or insight into the analyses presented in this report. A glossary of NFIRS terms relevant to these analyses is presented in Appendix A.

METHODOLOGY

Sources

This report is based on two different data sets: data on fire incidents from the National Fire Incident Reporting System (NFIRS) and mortality data from the National Center for Health Statistics (NCHS). Six years of mortality data (1983-1988) and three years of NFIRS data (1993-1995) serve as the basis for the findings in this report.

The National Fire Incident Reporting System (NFIRS) is an information system initiated and supported by the U.S. Fire Administration. The U.S. Fire Administration developed NFIRS as a means of assessing the nature and scope of the fire problem in the U.S. The system first came on line in 1976, and since then it has grown in both participation and use. Currently, nearly 14,000 of the nation's fire departments participate in the NFIRS and provide data on nearly one million fires each year. Not only does this make NFIRS the largest fire data set in the country, it also makes NFIRS the largest fire data set internationally.

The data sets chosen for the analysis of fire incidents are the United States Fire Administration's 1993, 1994, and 1995 NFIRS. Data from these three years have been averaged and are presented in the text and charts that appear in this report. The rationale behind analyzing three years of data lies in the benefit of evening out fluctuations that can arise in a single year's worth of data caused by an unusually cold winter season or other factors that could influence the frequency of fires. Additionally, three years is a narrow enough time period to avoid inadvertently smoothing out actual long-term trends in the fire data.

The National Center for Health Statistics (NCHS) mortality database is a compendium of deaths in the United States. The data is extracted from death certificates and the cause of death is coded in compliance with the International Classification of Diseases (ICD). Only a few of these codes apply to fire or fire-related mortalities. There is a small undercount of fire deaths associated with fire as fires in vehicles are classified as deaths due to vehicle accidents rather than fire. The NCHS data, until 1988, also included Federal Information Processing Standards (FIPS) codes in all data records. This information allows researchers to identify the county where individual deaths occurred and where decedents lived. After 1988, the smallest counties (those with populations fewer than 100,000 and for this analysis, the majority of the counties of interest) were no longer identified with FIPS codes. This makes it difficult, if not impossible, to identify deaths that occurred in rural areas. Hence, this analysis is based on NCHS data for years prior to 1988.

Defining Rural

There are many definitions of rural. Many, if not most, of the current "standard" definitions of rural provide a rationale for the appropriation and distribution of government funds. These definitions tend to differ from those used by demographers, economists, and others where small scale, low-density settlement and distance from large urban centers are the predominate considerations.

The federal government most frequently uses two classification schemes. These are the metropolitan/nonmetropolitan designation from the U.S. Office of Management and Budget (OMB), and the rural/urban designation from the U.S. Bureau of the Census. Both of these schemes define "rural" or "nonmetropolitan" areas not by what they are, but rather by what they are not. That is, "rural" areas are areas not considered "urban" or "metropolitan."

The metropolitan/nonmetropolitan classification from OMB is based on defined Metropolitan Statistical Areas (MSAs). MSAs consist of a central city of 50,000 or more residents, or an urbanized area of 100,000 or more population, and some of the area surrounding it. MSAs take into account the level of economic and social integration between the population center and the surrounding region. MSAs are usually specified at the county level and may consist of one county or a group of counties. The MSA classification can be problematic in that a county may be technically defined as part of an MSA but actually contain large expanses of land that are unquestionably rural.

The U.S. Census defines the rural population as those living in areas not categorized as urban. The urban population is those people living in an "urbanized area", or outside of an urbanized area in "places" with more than 2,500 residents; the rural population is then composed of those living outside of these areas.

Other classification schemes are also used to delineate urban and rural. The U.S. Department of Agriculture (USDA) uses a formula that includes a variety of indicators such as economic activity, county demographics, and geographical location with respect to a metropolitan area to classify each of the United States' 3,000-plus counties by "character of place." These characteristics are not mutually exclusive as any given county may meet the criteria for multiple categories. This method is based on a continuum rather than a binary (rural-urban) delineation. That is, there are degrees of "urbanness" or "ruralness."

For the purposes of this report, rural is defined using the USDA's Rural-Urban Continuum, commonly referred to as the Beale codes. The Beale codes were easily matched with the NCHS mortality data because both files include county Federal Information Processing Standards (FIPS) geographic codes. For NFIRS, the data were linked to the USDA data set by using fire department identification numbers in combination with FIPS codes. These processes allowed the identification of rural and non-rural fire deaths and fire incidents.

The full set of Beale codes is as follows:

Metropolitan counties:
0	Central counties of metropolitan areas of 1 million population or more
1	Fringe counties of metropolitan areas of 1 million population or more
2	Counties in metropolitan areas of 250,000 to 1 million population
3	Counties in metropolitan areas of fewer than 250,000 population

Non-metropolitan counties:
4	Urban population of 20,000 or more, adjacent to a metropolitan area
5	Urban population of 20,000 or more, not adjacent to a metropolitan area
6	Urban population of 2,500 to 19,999, adjacent to a metropolitan area
7	Urban population of 2,500 to 19,999, not adjacent to a metropolitan area
8	Completely rural or fewer than 2,500 urban population, adjacent to a metropolitan area
9	Completely rural or fewer than 2,500 urban population, not adjacent to a metropolitan area

For this analysis Beale categories 7, 8, and 9 are used as the working definition of rural. This definition includes those counties that are acknowledged to be completely rural along with those counties that have small urban population bases but are not adjacent to metropolitan areas. These are counties that probably have an "urban" county seat but do not have the nearby resources associated with being the neighbor of a larger metropolitan area.

According to the most recent Beale code data available (1993), 45.7 percent of all U.S. counties are considered to be "rural" under this definition. As presented in Table 1, below, these counties account for 19.4 million people or 7.5 percent of the estimated 260 million U.S. population in 1993.

**Table 1. Number of Counties and 1993 Population Distribution
by Beale Code**

Beale Code*	Total Counties	Total Population (in Millions)
7	654	13.2
8	248	2.6
9	530	3.6
Total "Rural"	1,432	19.4
Total U.S.	3,134	260.4

* Based upon 1993 Beale codes.

Unknowns and Adjusted Percentages

On a small portion of the incident reports submitted to NFIRS, the desired information for some or many data items is either left blank or reported as "unknown." To deal with this problem, this report cites adjusted percentages. This approach distributes fires with unknown characteristics in the same proportion as fires with known characteristics.

Representativeness of the Data

The NCHS data set is a census of deaths rather than a sample, so it is necessarily representative of the causes of all deaths. However, the NFIRS data set is a sample of all fire incidents that occur on an annual basis. Because NFIRS is a voluntary reporting system, participation in NFIRS varies state to state, and some states do not participate at all. In addition, NFIRS only includes fires to which the fire service was called. Nonetheless the distribution of participants is at least reasonably representative of the entire nation, though the sample is not random. On an annual basis, over 40 percent of all U.S. fires are reported to NFIRS; these are so well distributed geographically and by size of community that there is no known major bias that will affect the results of NFIRS analyses.

The NFIRS data on residential structure fire incidents track the distribution of the U.S. population very well – in fact the data on fires and population are remarkably consistent. Fires are often related to human activities, so the distribution of fires geographically would be expected to correspond to where people live. Using 1993 Census data, rural areas accounted for seven percent of the U.S. population. In the NFIRS data set, rural areas accounted for approximately eight percent of all residential structure fires. Conversely, non-rural areas account for 93 percent of the U.S. population and approximately 92 percent of all residential structure fires.

Rural fire departments are well represented in NFIRS. Fully one-fourth of all fire departments reporting to NFIRS are from rural areas. This over-representation among rural fire departments may be a result of the fact that a large number of rural fire departments are needed to attain reasonable response times in rural areas. Data from NFPA support this line of reasoning. Forty percent of all firefighters work or volunteer in areas with populations of fewer than 2,500.

Cause Categories

The cause of a fire is often the result of a complex chain of events. To make it easier to grasp the "big picture," 13 major categories of fire cause, such as heating and arson, are used by the U.S. Fire Administration. The cause categories used in the NFIRS portion of this project are listed in the same order on each graph to make comparisons easier from one to another.

Regional Breakdowns

To gain a fuller understanding of rural fire problems, the United States was divided into North and South regions and East and West regions (see Appendix B.) Few significant differences in the eastern and western regions were found in the NFIRS data. Where significant differences between the northern and southern regions were identified in the data, these findings are discussed in a special section of this report.

PART I. THE RURAL FIRE PROBLEM

Fire is a serious problem in the United States, more so than is generally perceived. Annually, there are millions of fires, thousands of deaths, tens of thousands of injuries, and billions of dollars in property loss – which together make the U.S. fire problem one of great national importance.[1] While the fire problem is evident everywhere, the nature of the problem is different depending on where you live. This report analyzes the unique characteristics of the fire problem in rural communities. While the distribution of fires in rural areas is strikingly similar to the U.S. fire problem as a whole, there are fundamental differences in the nature of the rural fire problem.

Where Fires Occur

Fires occur in the same types of locations in rural areas and in non-rural areas. Table 2 shows the distribution of rural fires and all U.S. fires broken down by property type. The table reveals that outside, structure, vehicle, and other fires occur in roughly the same proportion in rural areas as in the U.S. as a whole. The largest difference evident in Table 2 is for vehicle fires – 19 percent of rural fires are vehicle fires, compared to 24 percent of all U.S. fires.

Table 2. Distribution of Fires by Property Type

Property Type	Percent Distribution	
	Rural	*U.S.*
Outside Fires	45%	43%
Structure Fires	35%	31%
Non-residential Structure Fires	10%	9%
Residential Structure Fires	25%	23%
Vehicle Fires	19%	24%
Other Fires	1%	2%
TOTAL FIRES*	100%	100%

* may not add to 100 percent due to rounding

---

[1] "Fire in the United States 1985-1994." Washington, DC: Federal Emergency Management Agency, United States Fire Administration, National Fire Data Center. Ninth Edition.

Table 3 provides the distribution of casualties, both deaths and injuries, due to fire. Again, the distribution of deaths and injuries among outside, structure, vehicle, and other fires is very similar for rural areas and the entire U.S. There are slight differences, however. Outside and vehicle fires in rural areas account for slightly more casualties than for the U.S., while rural structure fires account for a slightly lower proportion of casualties.

Table 3. Distribution of Civilian Deaths and Injuries by Property Type

Property Type	Percent of Deaths		Percent of Injuries	
	Rural	U.S.	Rural	U.S.
Outside Fires	4%	3%	6%	5%
Structure Fires	73%	78%	76%	81%
Non-residential Structure Fires	4%	6%	15%	13%
Residential Structure Fires	69%	72%	60%	68%
Vehicle Fires	21%	17%	13%	10%
Other Fires	2%	3%	5%	4%
TOTAL FIRES*	100%	100%	100%	100%

* may not add to 100 percent due to rounding

While relatively fewer rural fires than non-rural fires occur in vehicles, rural vehicle fires account for a slightly higher proportion of fatalities and injuries than non-rural vehicle fires. This raises the possibility that vehicle fires can be more dangerous in rural areas because accidents may occur farther from fire and emergency medical service stations than in non-rural areas. As vehicle fires are most often the result of collision, a further analysis of rural vehicles is not pursued here.

Characteristics of the Rural Fire Problem

While the distribution of fires in rural areas is strikingly similar to the U.S. fire problem as a whole, there are fundamental differences in the nature of the rural fire problem. Among these differences are the causes of rural fires and the characteristics of residential structure fires, including how they start, where they originate, and how often smoke detectors are present and in operational condition.

Outside Fires

Outside fires account for 45 percent of all fires that occur in rural areas. Although these fires represent a high proportion of all rural fires, they are rarely associated with fire deaths or injuries. An average of only four percent of rural fire deaths and six percent of rural fire injuries were associated with outside fires from 1993-1995.

The major causes of outside fires in rural and non-rural areas are profiled in Figure 1. The top three causes of outside fires reported to rural fire departments are open flame (45 percent), arson (16 percent), and natural causes (nine percent). In contrast to rural areas, in non-rural areas arson is the leading cause of outside fires and, at 44 percent, is nearly three times greater a problem than in rural areas.

To further investigate patterns in outside rural fires separate North/South and East/West analyses were conducted (see Appendix B for a list of states by region). There were no significant differences in outside rural fires for the North/South analysis, but the East/West analysis suggested some differences in the relative frequency and causes of outside fires.

One significant difference between the East and the West is the proportion of rural fires that occur outside. In the West, outside fires account for 55 percent of all rural fires. In the East, only 36 percent of all rural fires are outside fires. This difference may stem from the relative difference in land use in the East and West. With larger tracts of open land in the western U.S., a higher proportion of outside fires may be expected.

The leading cause of outside fires is the same in the East and the West – open flame accounts for over 40 percent of outside fires in both regions. While the second leading cause, arson, is also the same in both regions, it represents a much greater problem in the East. Arson accounts for fully 29 percent of outside fires in rural areas of the East, compared to 12 percent in the West (Figure 2).

Structure Fires

When people think about fire, they generally think of fires that occur in buildings. Thirty-five percent of all rural fires occur in structures, a slightly higher proportion than is observed for the U.S. as a whole. This could be a cause for concern since structure fires, and residential structure fires in particular, are responsible for the vast majority of

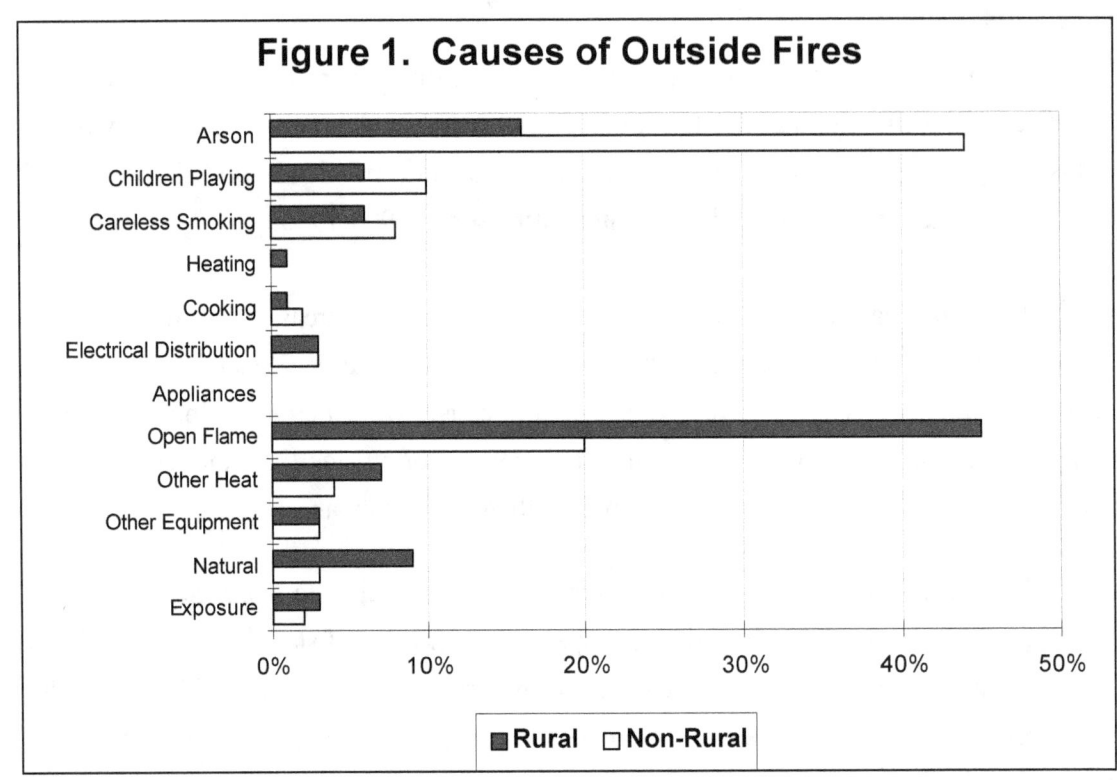

Figure 1. Causes of Outside Fires

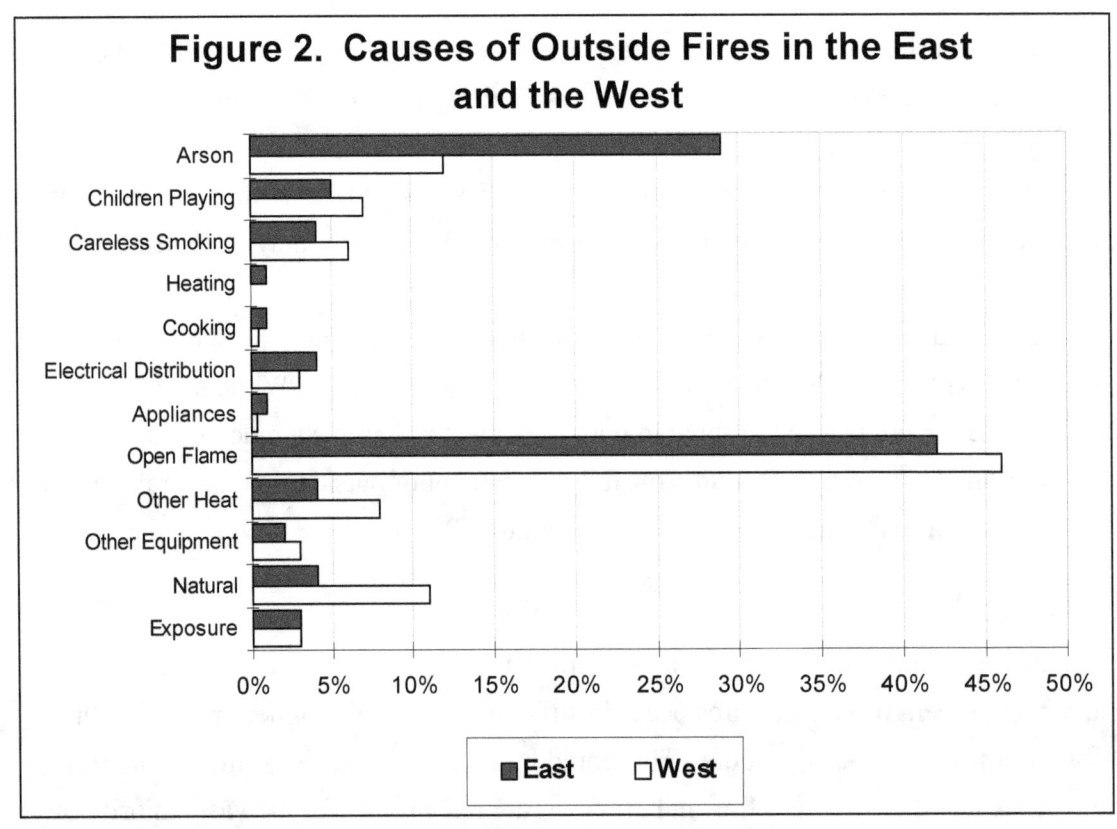

Figure 2. Causes of Outside Fires in the East and the West

Data Source: 1993 - 1995 National Fire Incident Reporting System

civilian deaths and injuries due to fire. However, as noted earlier, structure fires in rural areas account for a slightly lower proportion of fire deaths and injuries than is the case for the entire U.S.

Figure 3 illustrates the three leading causes of structure fires in rural areas. These are heating (29 percent), arson, and electrical distribution (tied at 12 percent). In contrast, the three leading causes of structure fires in non-rural areas are cooking, arson, and heating.

Figure 4 indicates that the three leading causes of fatal fires in rural structures are heating (25 percent), careless smoking (21 percent), and electrical distribution (17 percent). This is in stark contrast to the rank order of leading causes of fatal fires in non-rural areas. Careless smoking is the leading cause of non-rural fatal fires (27 percent), followed by arson (18 percent), and heating (12 percent).

To fully understand the nature of structure fires in rural areas, it is useful to analyze non-residential and residential structures separately. Non-residential structures are discussed below, while a separate section of this report is devoted to residential structures, where the vast majority of fire deaths and injuries occur.

Non-Residential Structure Fires. Fires in non-residential structures make up 10 percent of rural structure fires. While these fires account for only four percent of deaths, they account for 15 percent of injuries from rural fires. The leading causes of rural non-residential structure fires are arson and open flame, which each account for 18 percent of fires (Figure 5). Electrical distribution is the third leading cause, accounting for 13 percent of fires. In contrast, arson is by far the leading cause of non-residential structure fires in non-rural areas. This cause alone accounts for 31 percent of these non-rural fires. Electrical distribution is the second leading cause in non-rural areas, accounting for 11 percent of all non-residential structure fires, and open flame is third, accounting for 10 percent of fires.

Analyzing the causes of fatal fires in rural non-residential structures reveals that heating, electrical distribution, open flame, and natural causes are tied for the leading

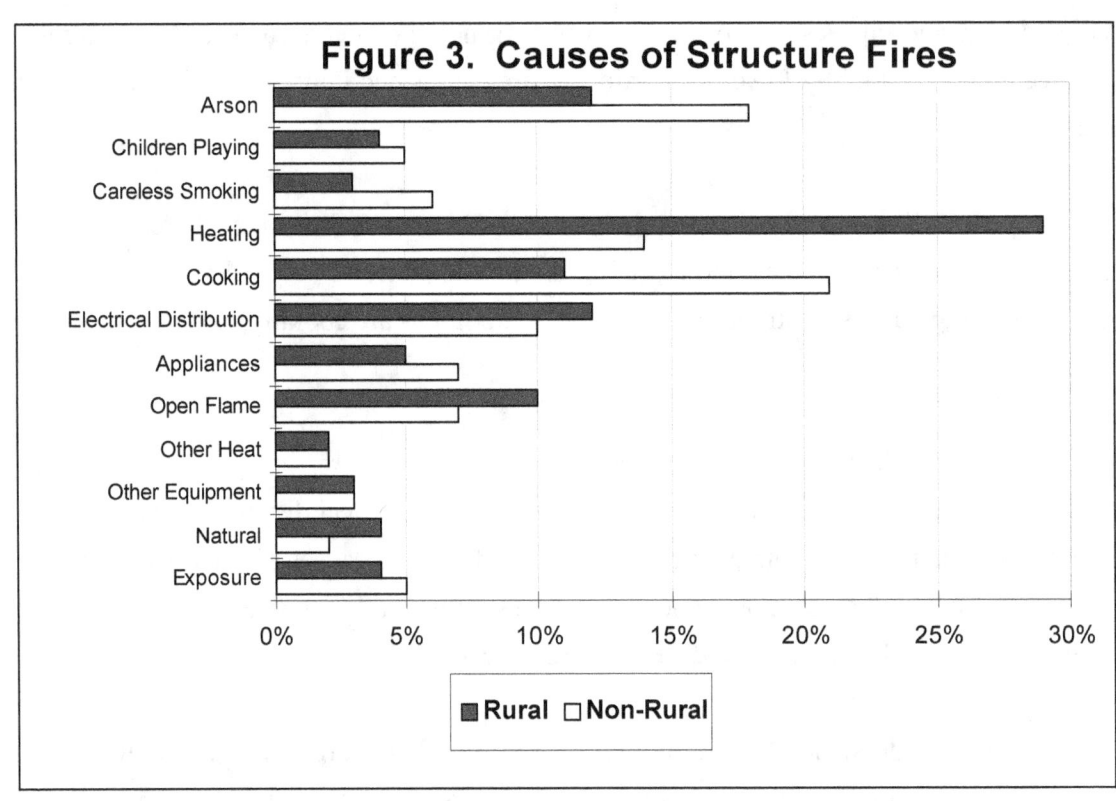

Figure 3. Causes of Structure Fires

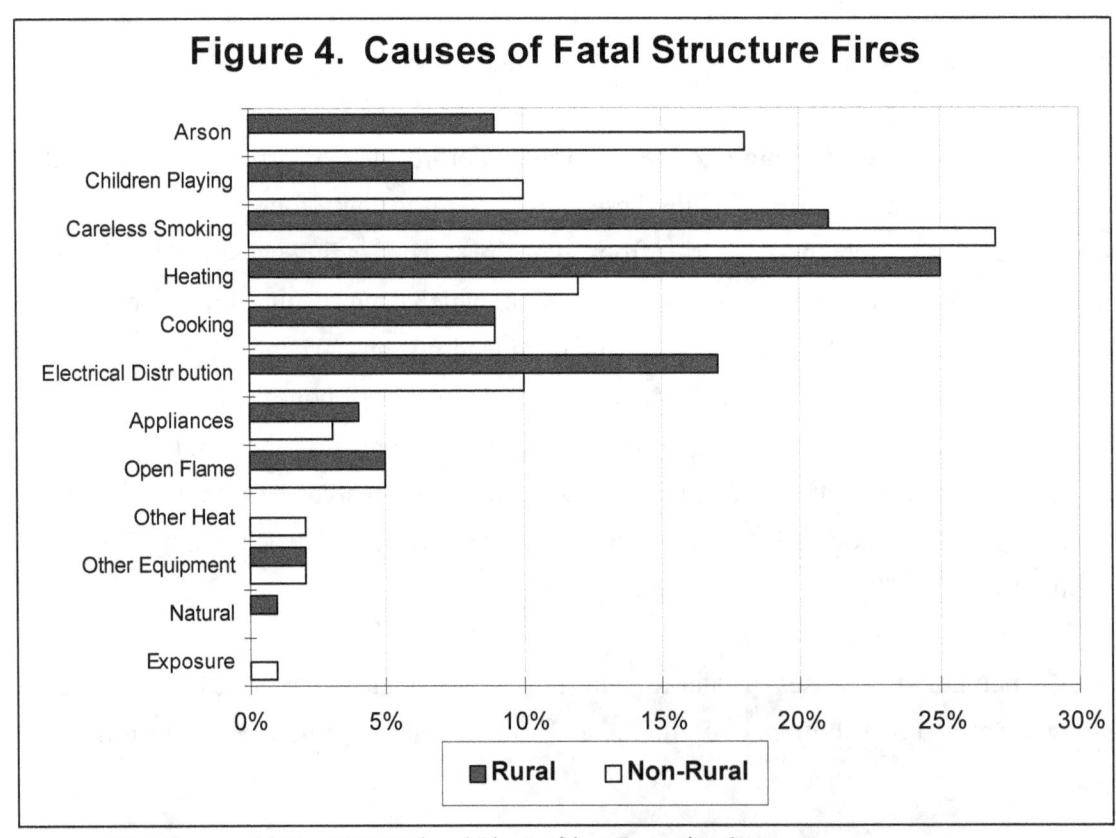

Figure 4. Causes of Fatal Structure Fires

Data Source: 1993 - 1995 National Fire Incident Reporting System

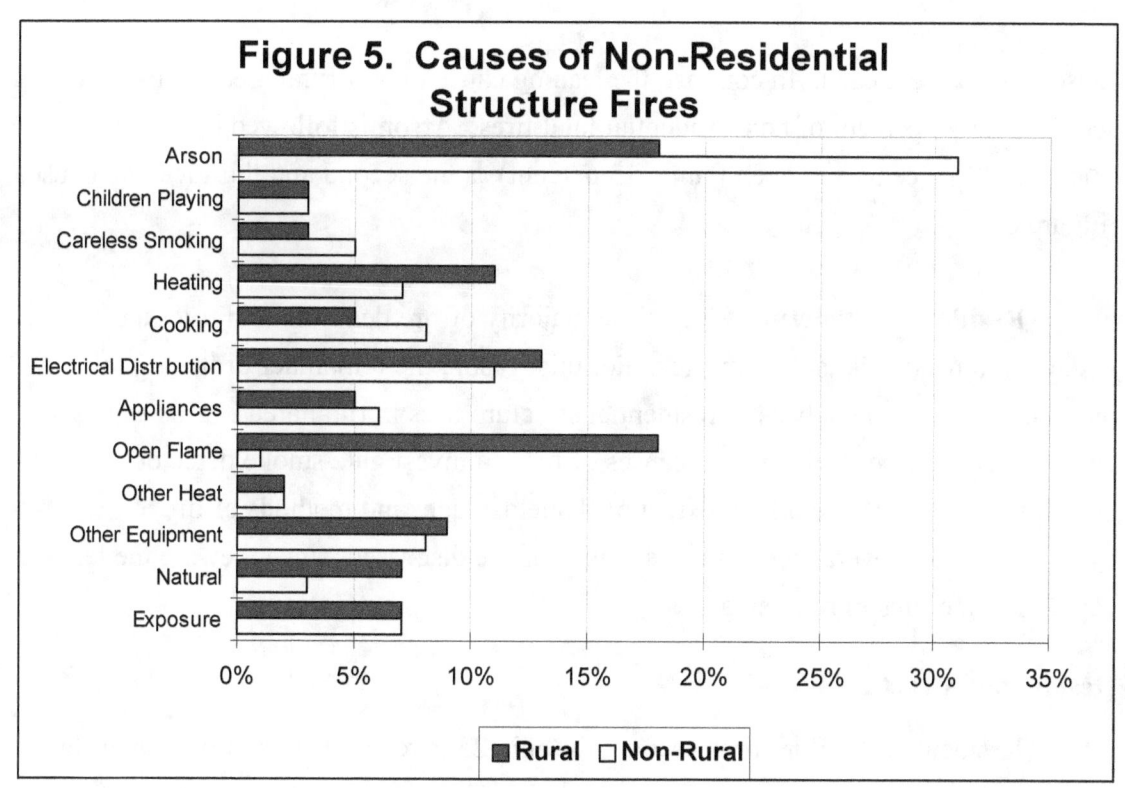

Figure 5. Causes of Non-Residential Structure Fires

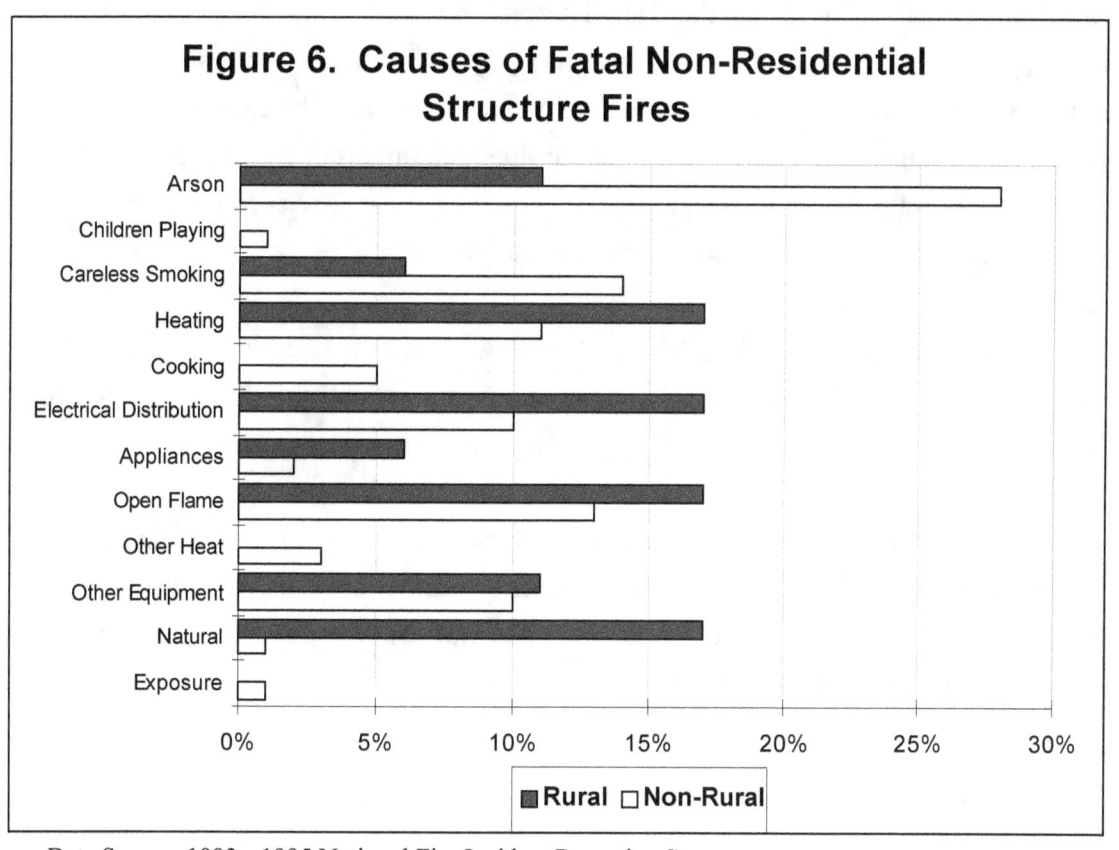

Figure 6. Causes of Fatal Non-Residential Structure Fires

Data Source: 1993 - 1995 National Fire Incident Reporting System

cause at 17 percent each. In contrast, the leading cause in non-rural areas is arson, which accounts for 25 percent of non-residential fatal fires. Arson is followed by careless smoking (14 percent) and open flame (13 percent) as the second and third leading causes (Figure 6).

Residential Structure Fires. The majority of fire deaths and injuries are sustained in residential structure fires. For this reason, the remainder of this report provides an in-depth analysis of residential structure fires in rural areas. This analysis extends beyond an examination of causes of fires to investigate smoke detector performance, area of fire origin, extent of flame damage, and methods of fire extinguishment. Several aspects of heating fires are discussed, since they are the leading cause of home fires in rural areas.

Residential Structure Fires

Residential structure fires account for only 25 percent of fires in rural areas but over two-thirds of fire deaths and 60 percent of fire injuries. The leading cause of rural residential structure fires is heating – at 36 percent it is nearly three times as prevalent as the second leading cause, cooking (13 percent) (Figure 7). Moreover, residential structure fires in rural areas are more than two times as likely to be caused by heating as fires in non-rural areas. This finding is of considerable importance because of the higher incidence of deaths and injuries in residential structure fires. The specific characteristics of rural heating fires are discussed in greater detail below.

After heating and cooking, the next leading cause of residential structure fires in rural areas is electrical distribution (12 percent). Arson is the fourth leading cause, but it is slightly less severe of a problem in rural areas than in non-rural areas.

In non-rural areas of the U.S., cooking rather than heating is the leading cause of residential structure fires and causes one-quarter of these fires. Heating is the second leading cause, accounting for 16 percent of residential structure fires, and arson is third, accounting for 14 percent of fires.

Fatal Fires

Figure 8 shows that the top three causes of fatal fires in rural homes are heating (26 percent), careless smoking (23 percent), and electrical distribution (17 percent). In contrast, heating fires account for only 12 percent of fatal fires in non-rural areas, making them the third leading cause rather than the leading cause. In non-rural areas the top two leading causes are careless smoking (28 percent) and arson (17 percent).

Fires with Injuries

The leading causes of fires with injuries in rural homes are heating and cooking, tied at 23 percent (Figure 9). Children playing, careless smoking, and electrical distribution are tied as the next most common causes, each accounting for 10 percent of fires with injuries. In non-rural areas, cooking is the leading cause (30 percent) and out-distances all other causes by a significant margin. The next two leading causes are careless smoking and children playing fires, each accounting for 12 percent of fires with injuries.

Smoke Detector Performance

The majority (58 percent) of rural fires occur in homes without smoke detectors, as is shown in Figure 10. To compound this problem, 15 percent of rural fires occur in homes where smoke detectors are present but do not operate. Therefore 73 percent of rural residential structure fires occur in homes without *operational* smoke detectors. This is a very alarming fact. Since this study uses "fires" as its unit of analysis, additional research is needed to investigate whether a lower proportion of rural homes *in general* lack smoke detectors.

Figure 11 shows that the pattern is somewhat different among non-rural fires. Fifty-eight percent of fires in non-rural areas occur in residential structures that have smoke detectors present. However, a larger proportion of non-rural homes experiencing fires has smoke detectors that are not operational. Thus, 65 percent of non-rural homes experiencing fires do not have the safety protection of operating smoke detectors.

One possible explanation for the higher incidence of rural residential fires in homes without smoke detectors is that residents of rural areas may have less access to public safety campaigns and may be less aware of the need for smoke detectors.

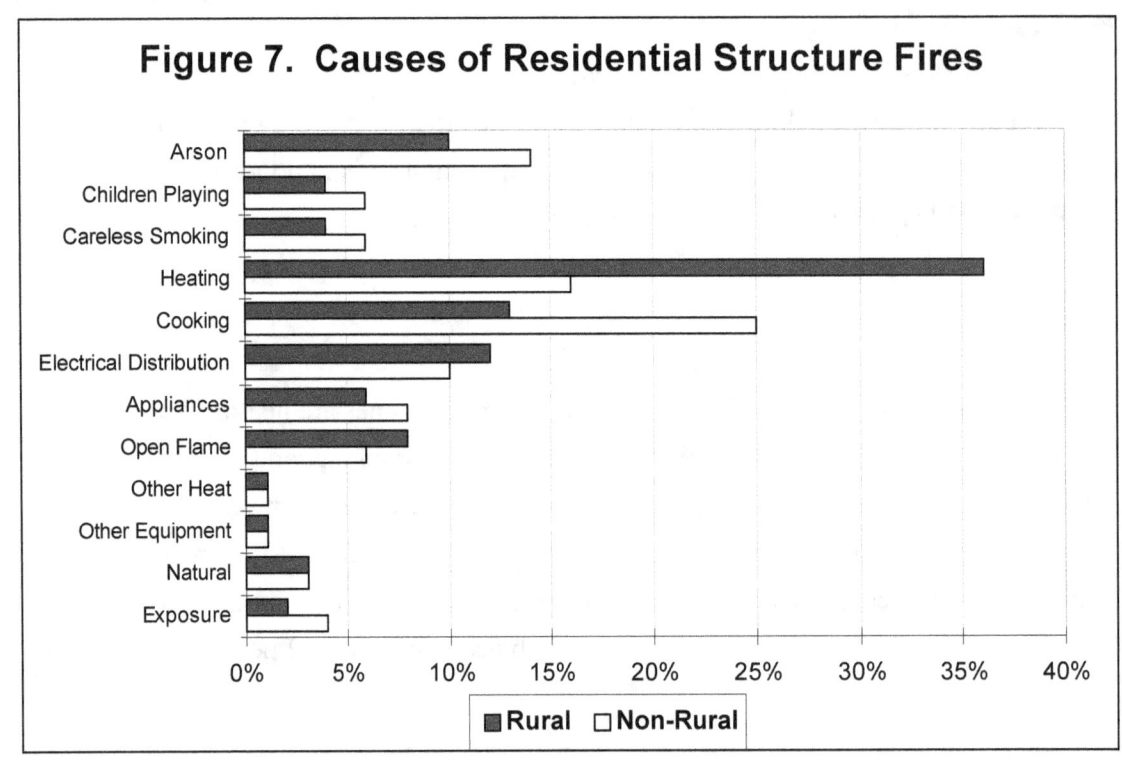

Figure 7. Causes of Residential Structure Fires

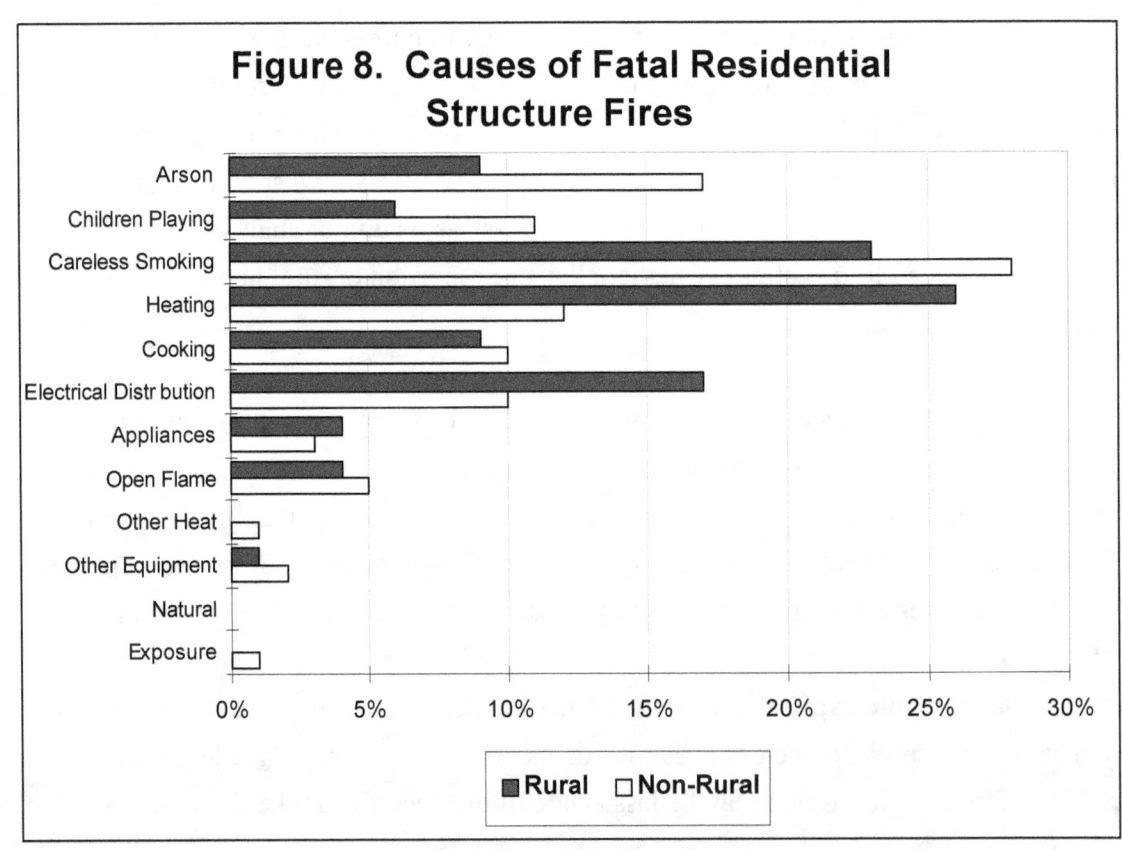

Figure 8. Causes of Fatal Residential Structure Fires

Data Source: 1993 - 1995 National Fire Incident Reporting System

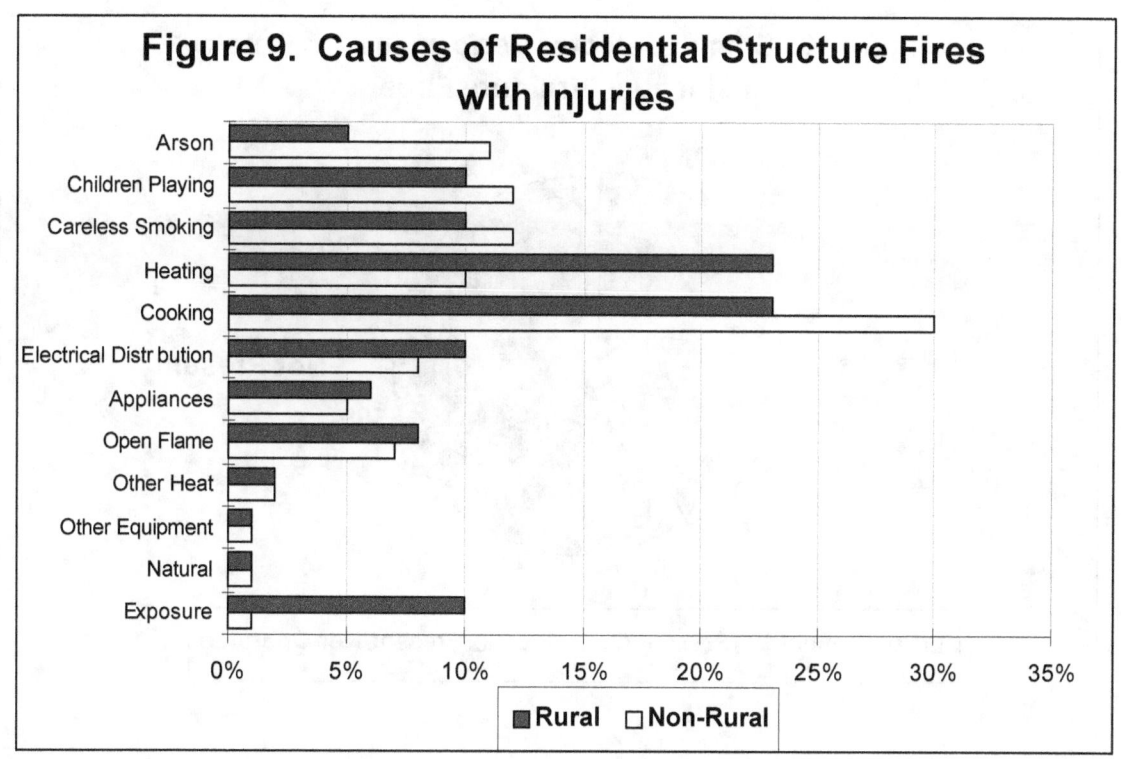

Figure 9. Causes of Residential Structure Fires with Injuries

Data Source: 1993 - 1995 National Fire Incident Reporting System

17

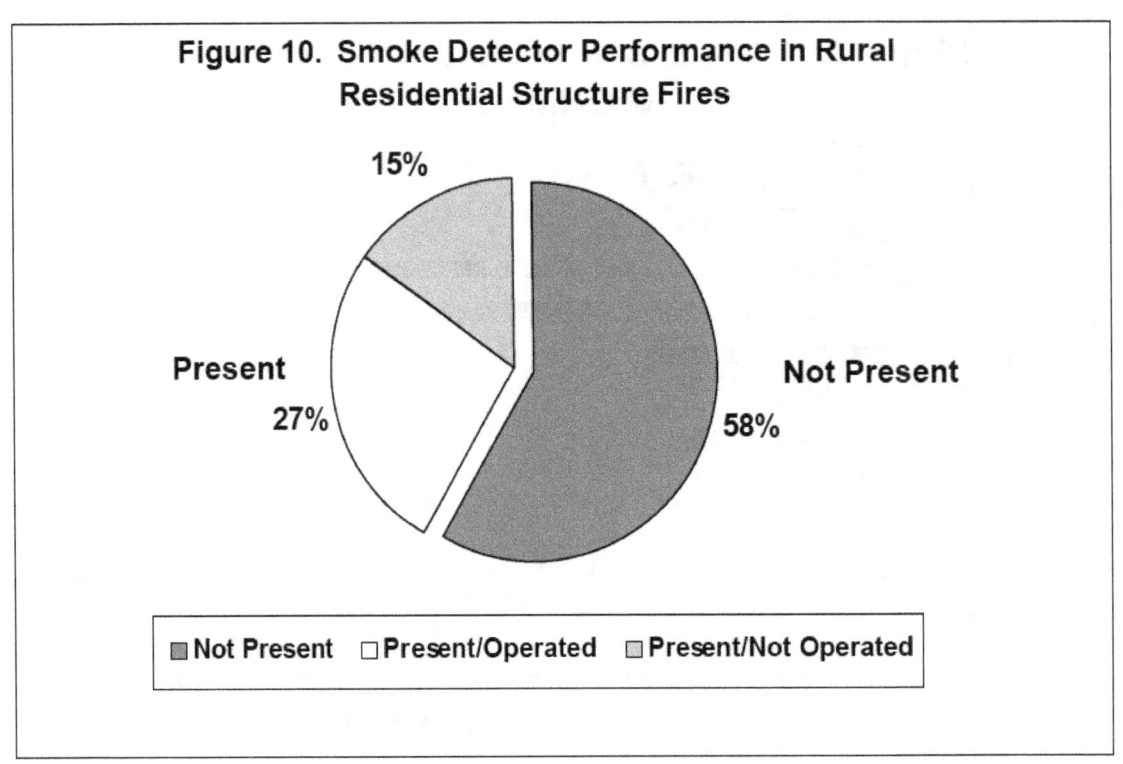

Figure 10. Smoke Detector Performance in Rural Residential Structure Fires

15%

Present
27%

Not Present
58%

■ Not Present   ☐ Present/Operated   ■ Present/Not Operated

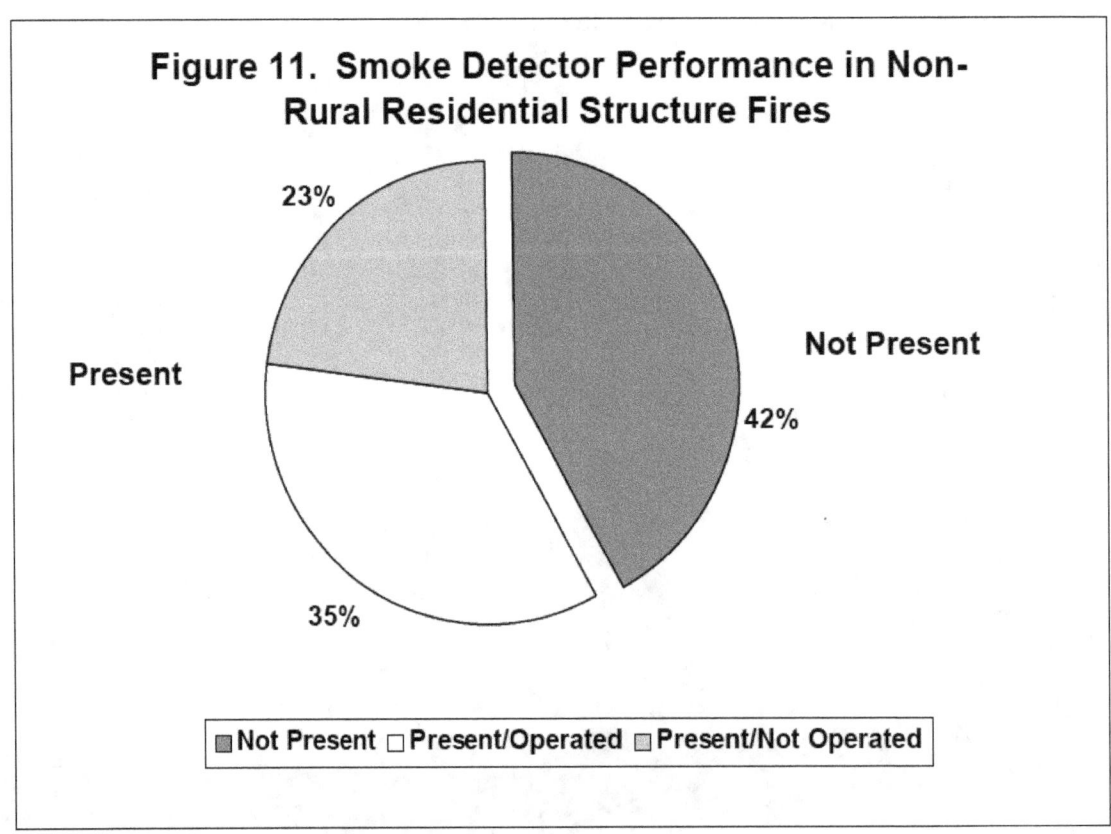

Figure 11. Smoke Detector Performance in Non-Rural Residential Structure Fires

23%

Present

Not Present
42%

35%

■ Not Present   ☐ Present/Operated   ■ Present/Not Operated

Data Source: 1993 - 1995 National Fire Incident Reporting System

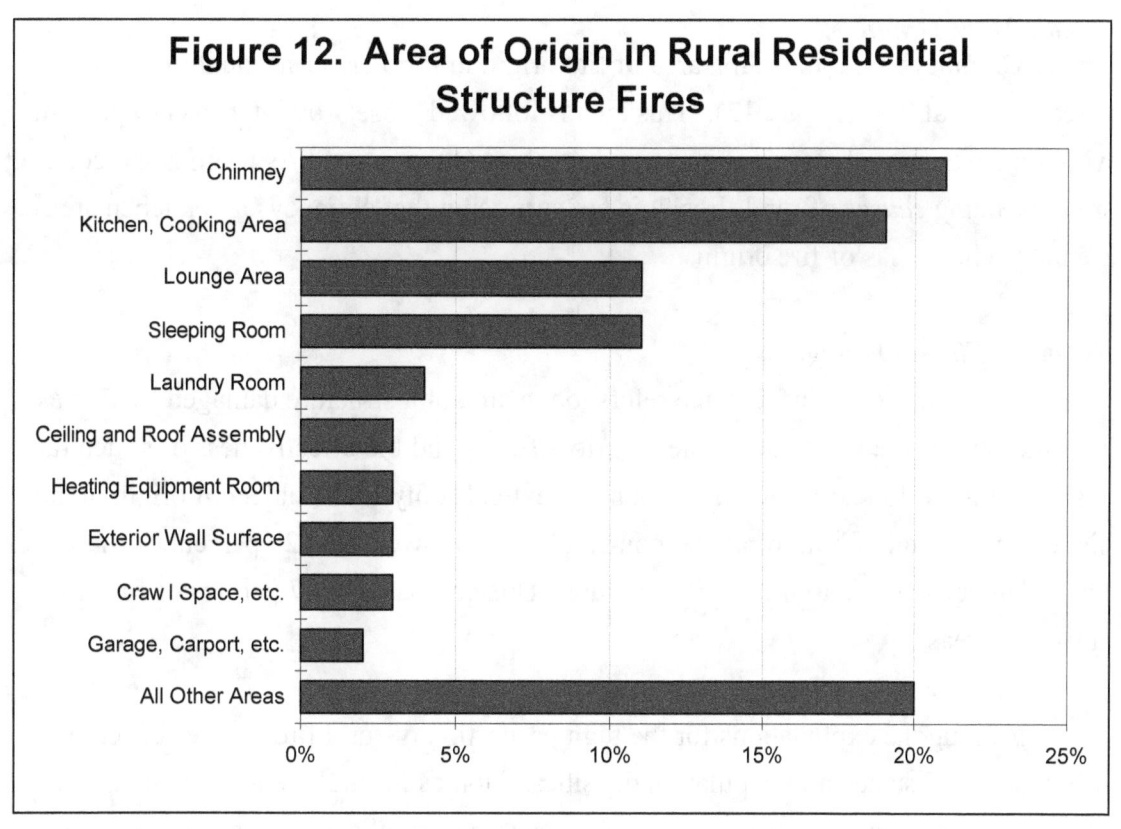

Figure 12. Area of Origin in Rural Residential Structure Fires

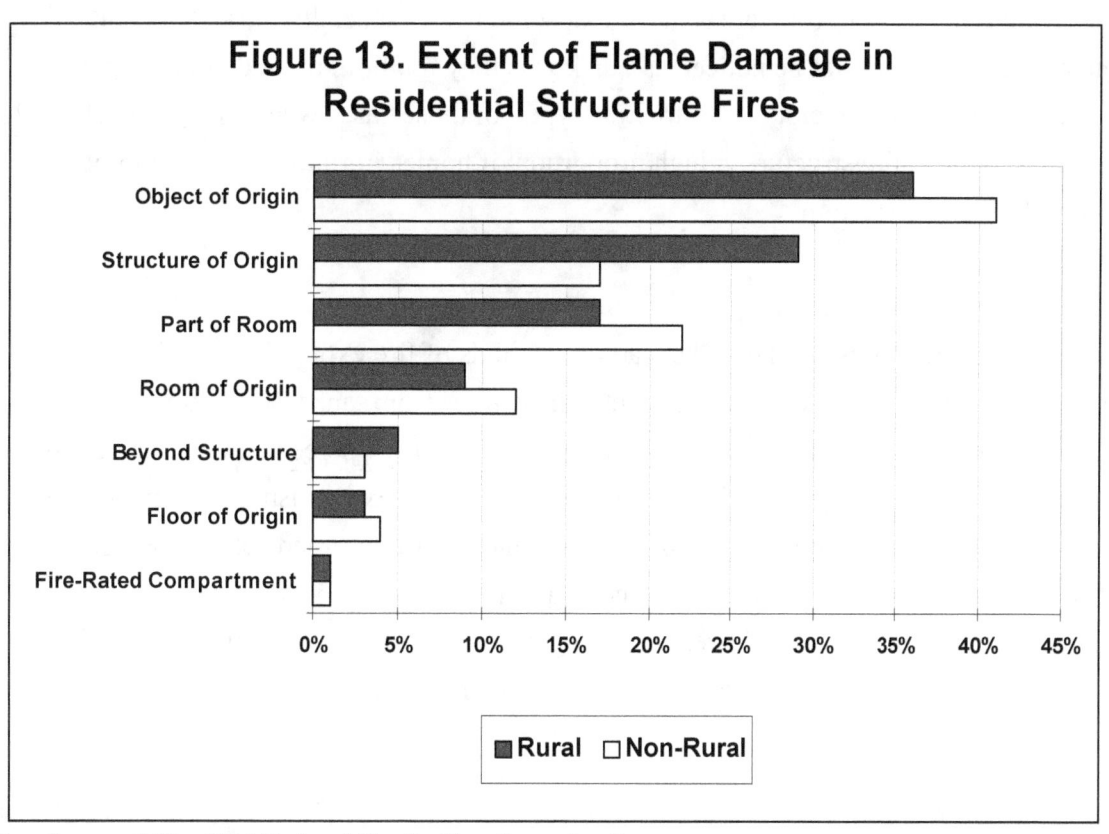

Figure 13. Extent of Flame Damage in Residential Structure Fires

Data Source: 1993 - 1995 National Fire Incident Reporting System

Area of Fire Origin

Chimneys are the leading area of fire origin in rural areas, accounting for 21 percent of rural fires (Figure 12). This area is followed closely by kitchens (19 percent), with lounge areas and sleeping rooms tied for third (11 percent). As heating and cooking are the leading *causes* of rural fires, it is not surprising that chimneys and kitchen areas are the leading areas of fire origin.

Extent of Flame Damage

The extent of flame damage refers to the area of a structure damaged by flames. This measure is used to evaluate the severity of fires and the effectiveness of structural designs. Figure 13 shows that flame damage extends only to the object of origin in the highest proportion of both rural and non-rural fires. However, in 29 percent of rural fires, flame damage extends to the entire structure. This compares to 17 percent of fires in non-rural areas.

Among the explanations for the high proportion of rural fires that effect entire structures are distance and population densities. Homes in rural areas are more likely to be farther from emergency services than homes in non-rural areas. Also, lower population densities in rural areas mean that some fires, especially those that begin while no one is home, will not be noticed as quickly as they might in more densely populated areas. An additional factor is construction type. In those cases where flame damage extends to the entire structure, a high proportion of homes in rural areas are unprotected wood frame construction.

Method of Extinguishment

Figure 14 illustrates that the leading methods of fire extinguishment in rural residential structure fires are pre-connected hose lines from tanks, self-extinguishing fires, and make-shift aids. These account for 46 percent, 14 percent, and 12 percent of fires, respectively. Similarly, the leading methods of fire extinguishment in non-rural areas are pre-connected hose lines from tanks, make-shift aids, and self-extinguishing fires, accounting for 30 percent, 19 percent, and 17 percent of fires, respectively. Nine percent of all rural residential fires are extinguished by portable extinguishers. This is thirty percent less often than occurs in residential structures in non-rural areas.

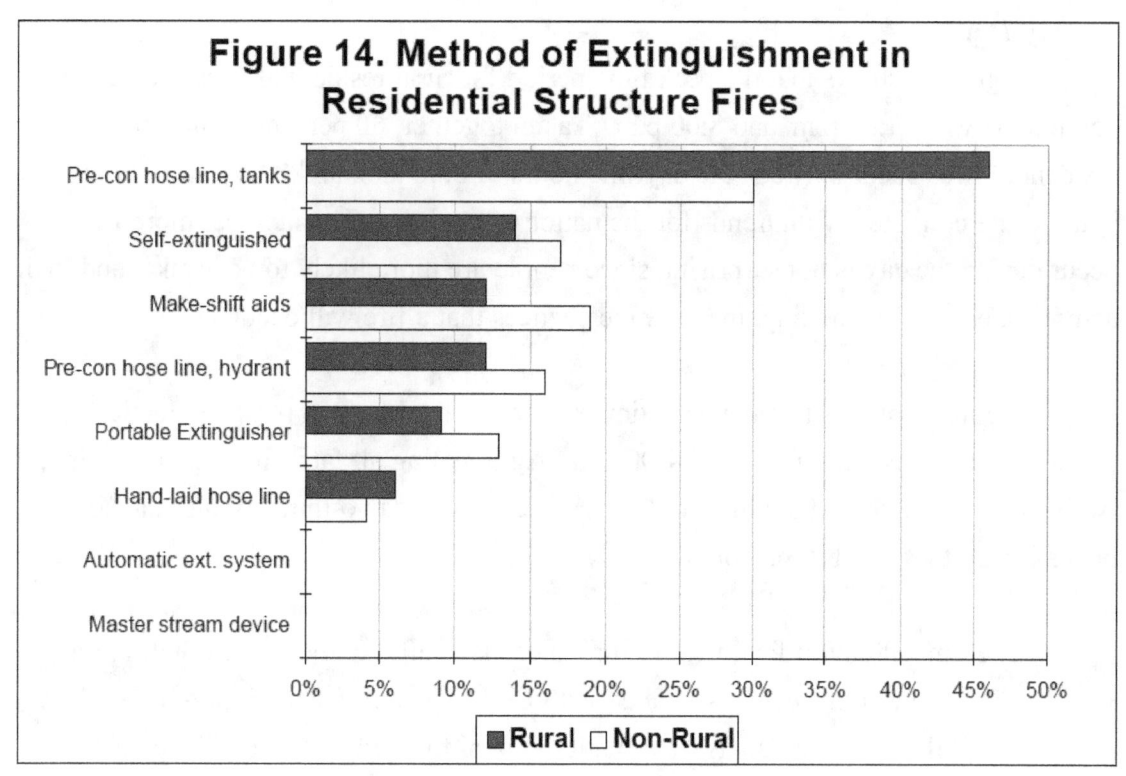

Figure 14. Method of Extinguishment in Residential Structure Fires

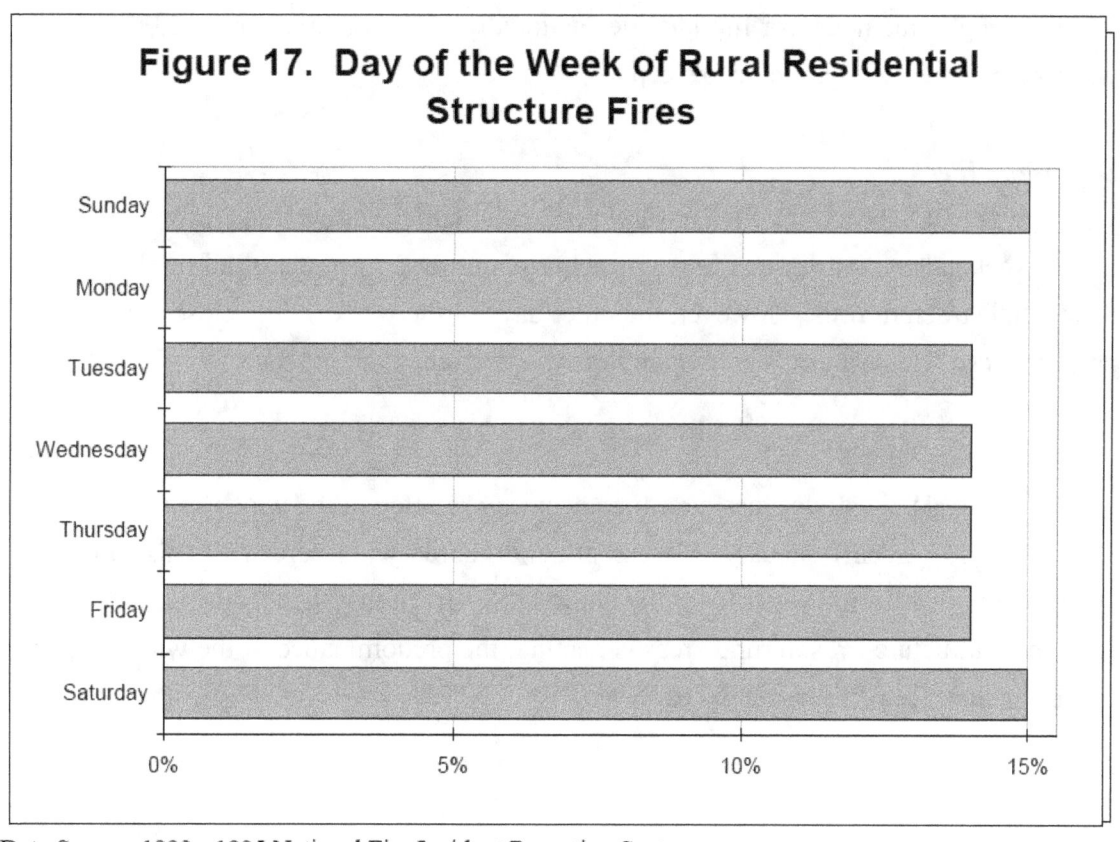

Figure 17. Day of the Week of Rural Residential Structure Fires

Data Source: 1993 - 1995 National Fire Incident Reporting System

21

Time of Day

Figure 15 shows that the peak time period for rural residential fires is during the daytime between 4:00 p.m. and 8:00 p.m. Taken together, 60 percent of all rural residential fires occur between the daytime hours of 8:00 a.m. and 8:00 p.m. These patterns are consistent with trends for the nation as a whole. The fact that more fires occur during the day is not surprising since people are more likely to be awake, and their activities, especially cooking, increase the chances that a fire will occur.

Figure 16 shows that the peak time period for fatal fires in rural residential structures is between midnight and 4:00 a.m. Aggregating all fatal fires that occur in the twelve hours between 8:00 p.m. and 8:00 a.m. reveals that two-thirds of all fatal fires occur during these nighttime hours.

The time of day of fatal fires in rural areas is similar to the United States as a whole. Fire deaths are usually associated with fires that start late at night or early in the morning. While equivalent analyses of national fatal fires are not available, over 60 percent of all fire fatalities occur between 8:00 in the evening and 8:00 the next morning. The peak nighttime hours for fire fatalities nationally are from 3:00 a.m. to 5:00 a.m., when people are getting into deep sleep.

Day of the Week

Figure 17 shows that fires in rural residential structures occur slightly more often on the weekends. Saturday and Sunday each account for 15 percent of all rural residential structure fires. However, the other days of the week follow closely, at 14 percent each. This pattern is consistent with trends at the national level.

Month of the Year

Figure 18 illustrates that rural fires occur more often in the winter months of December (14 percent), January (12 percent), and February (11 percent). This pattern is consistent with seasonal patterns for the entire U.S. Given that the leading cause of residential structure fires in rural areas is heating, the predominance of the winter months for the occurrence of fires is to be expected.

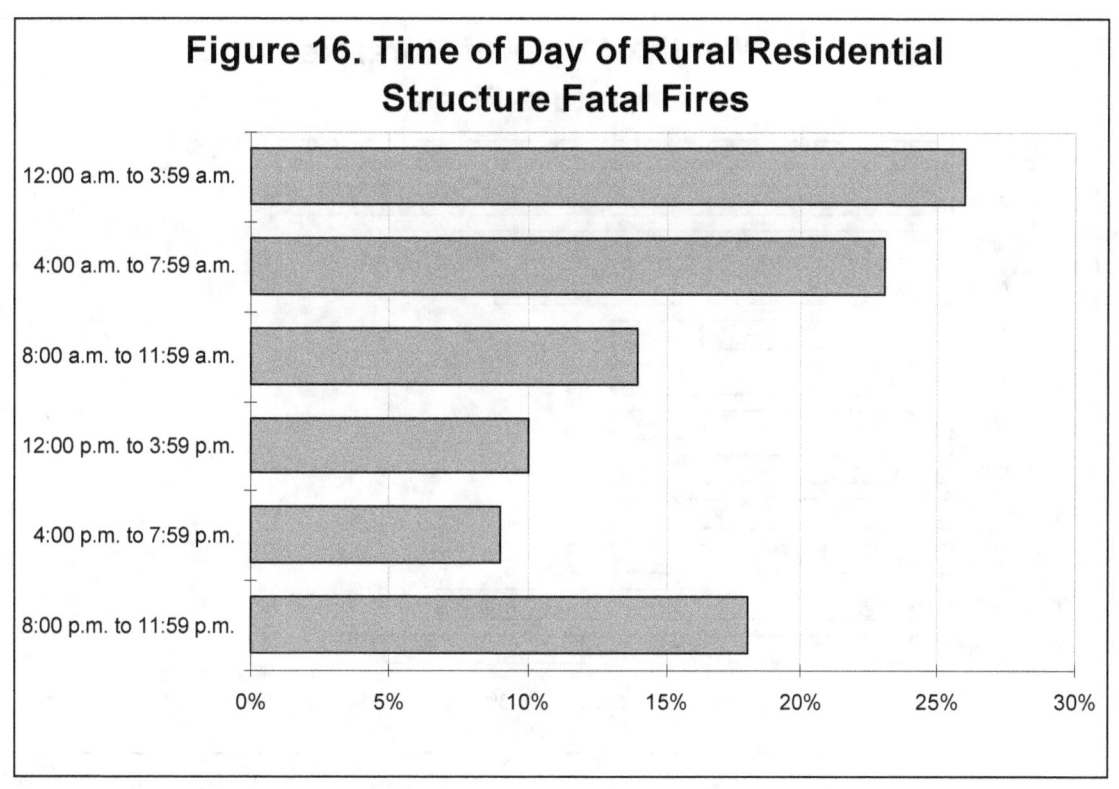

Figure 16. Time of Day of Rural Residential Structure Fatal Fires

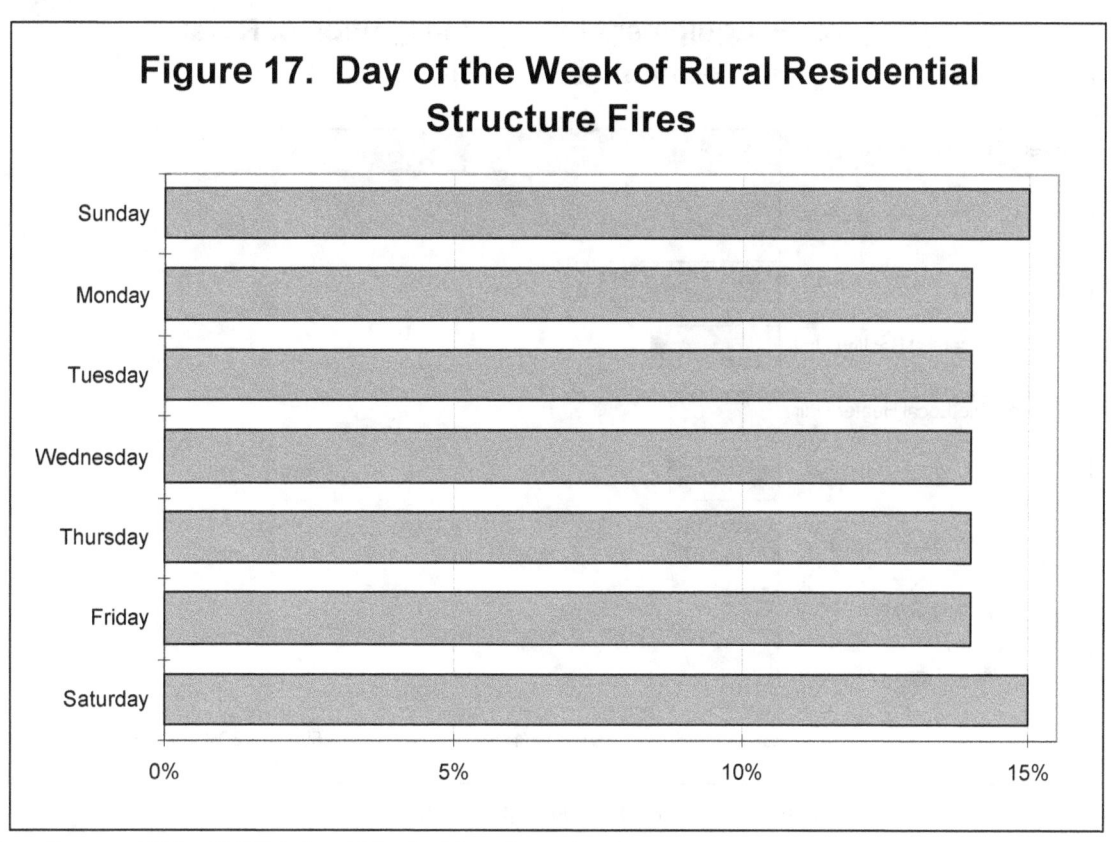

Figure 17. Day of the Week of Rural Residential Structure Fires

Data Source: 1993 - 1995 National Fire Incident Reporting System

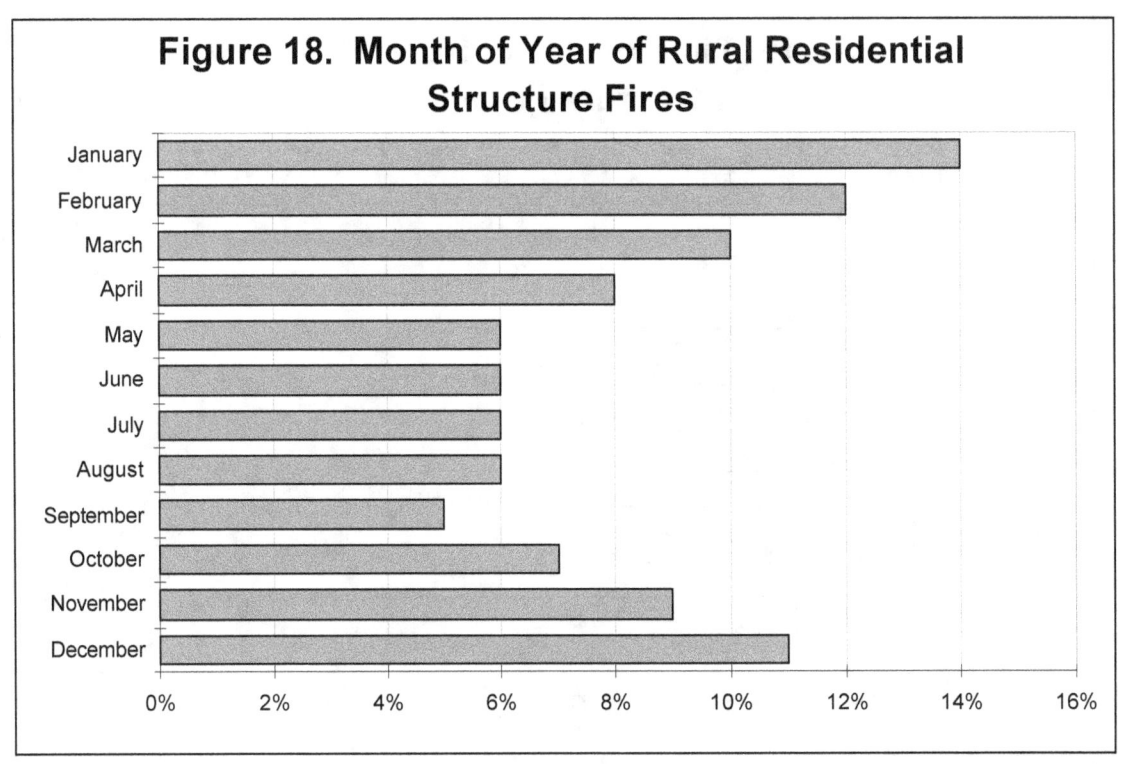

Figure 18. Month of Year of Rural Residential Structure Fires

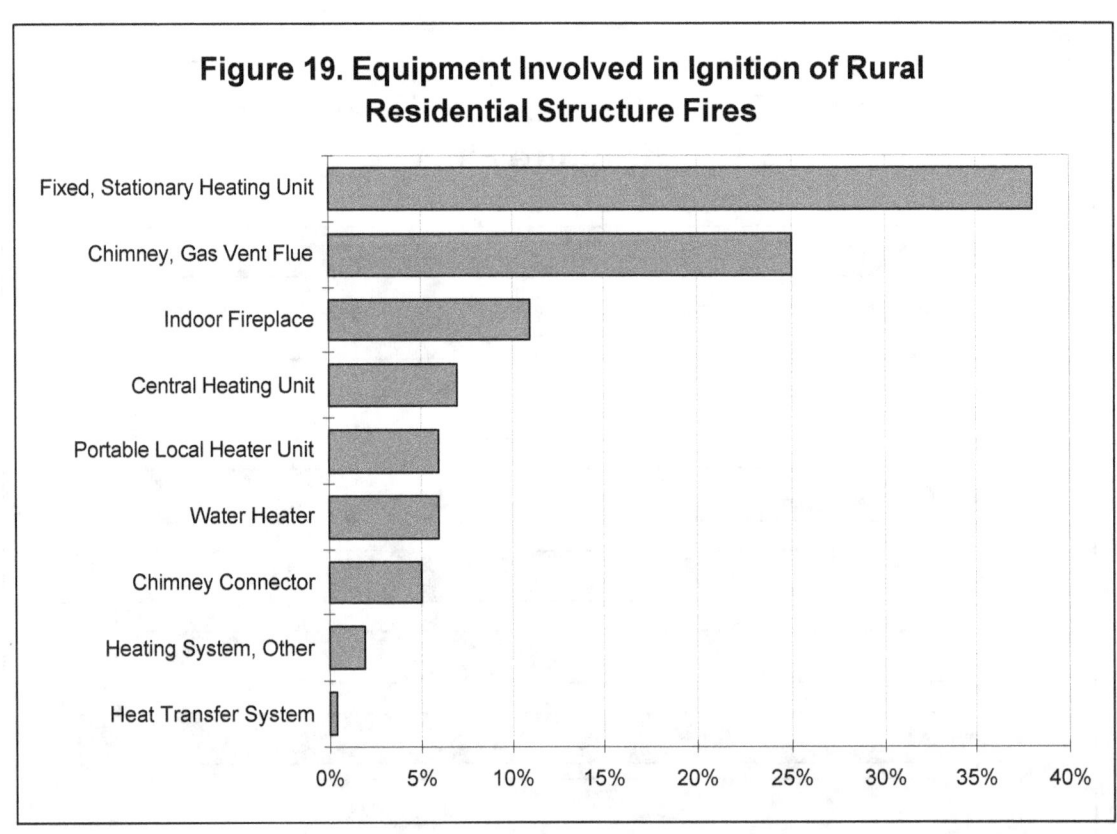

Figure 19. Equipment Involved in Ignition of Rural Residential Structure Fires

Data Source: 1993 - 1995 National Fire Incident Reporting System

Heating Fires

Because heating fires are central to an understanding of the fire problem in rural areas, this section is devoted to a more in-depth investigation of these fires. This includes discussions of the equipment involved in the ignition of rural heating fires, the types of materials most commonly ignited, and the most common ignition factors.

Equipment Involved in Ignition. Figure 19 indicates that the leading equipment involved in fire ignition for rural heating fires is fixed stationary heating units, accounting for 38 percent of all residential heating fires. Fixed stationary heating units include equipment such as woods stoves and kerosene heaters that are maintained in one location rather than moved around. Chimneys and indoor fireplaces, accounting for 25 percent and 11 percent of fires respectively, are the second and third leading pieces of equipment involved in ignition.

Type of Material Ignited. Consistent with the fact that many rural heating fires start in fixed stationary heating units, chimneys, and fireplaces, the leading category of the type of material ignited is adhesive, resin, and tar (Figure 20). Resin is the material that builds up in chimneys and is often responsible for chimney fires. Adhesive, resin, and tar are the types of material ignited in almost half of all heating fires in rural areas. The next most common type of material ignited is sawn wood (19 percent), another material associated with wood stoves and fireplaces.

Ignition Factor. Because of the preponderance of rural heating fires that originate in fixed stationary heating units, chimneys, and indoor fireplaces, a special analysis of the ignition factors for these particular fires was conducted. It revealed that the leading ignition factor is mechanical failure or malfunction, which accounts for 64 percent of these fires (Figure 21). Misuse of material ignited and operational deficiency, accounting for 11 percent and 10 percent respectively, are the second and third leading ignition factors. One important finding of this report is that people need to be educated about the proper maintenance and use of various heating devices, and they need to know that their chimneys need to be professionally cleaned and inspected each year before they use them.

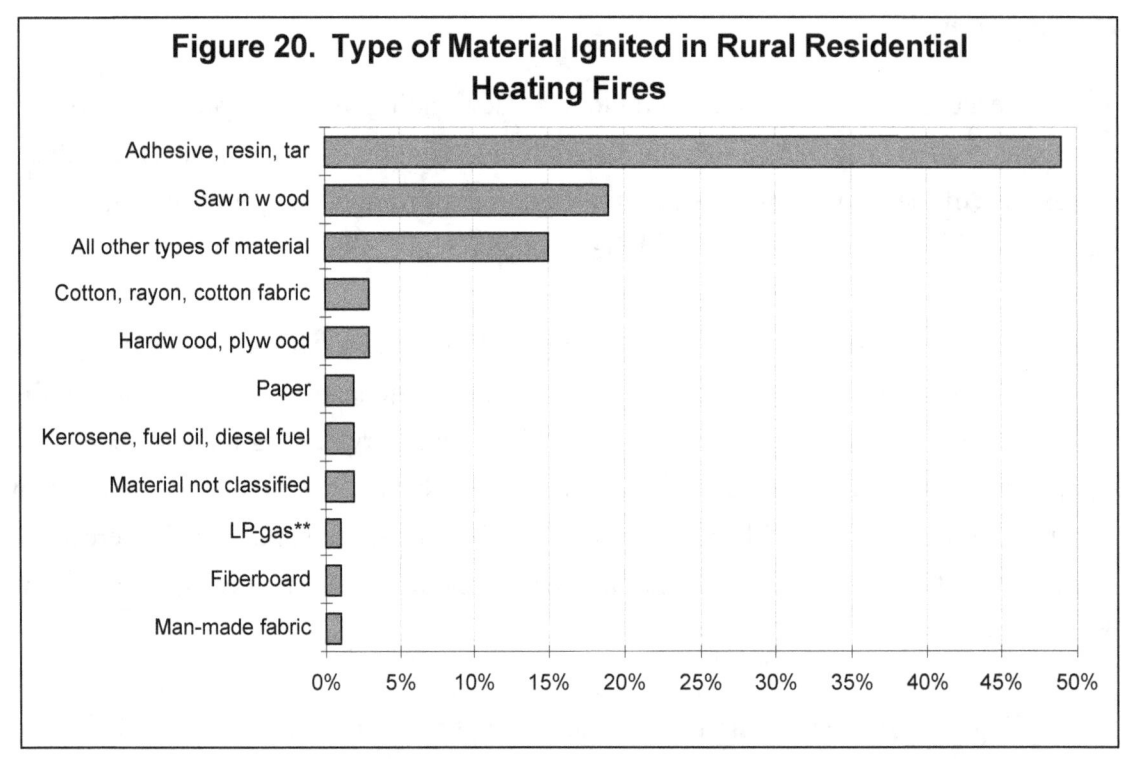

Figure 20. Type of Material Ignited in Rural Residential Heating Fires

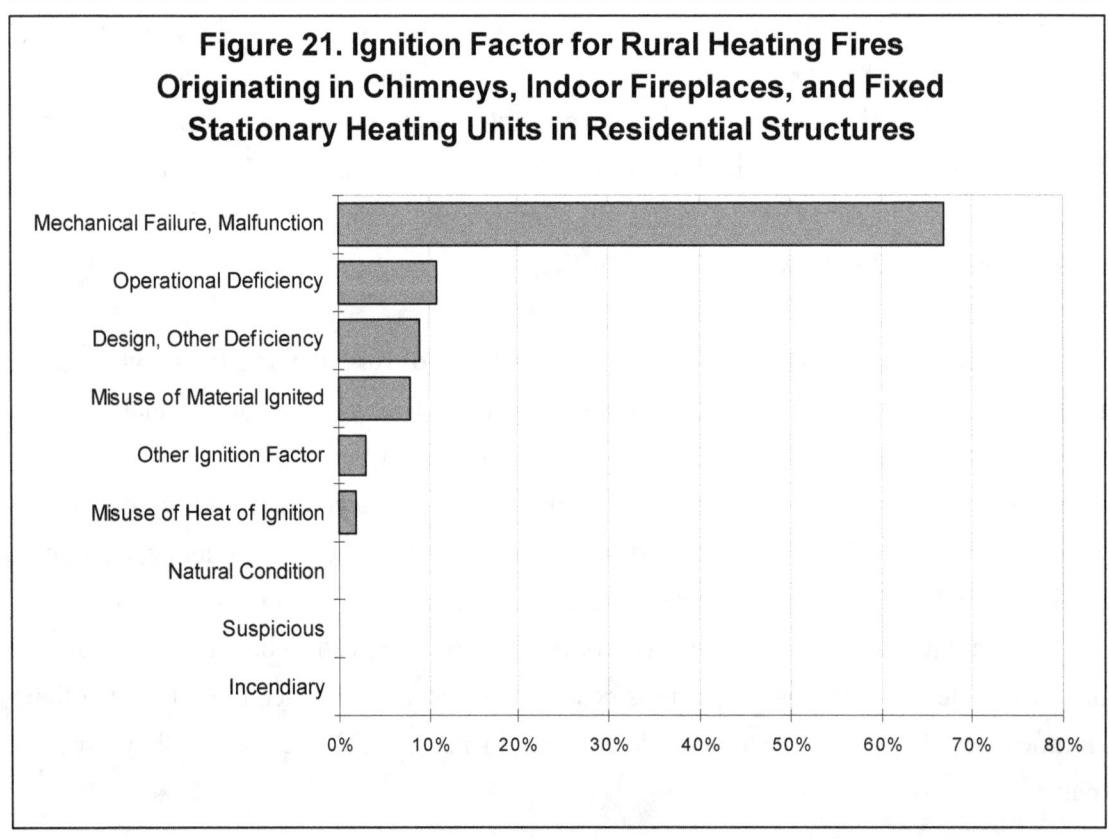

Figure 21. Ignition Factor for Rural Heating Fires Originating in Chimneys, Indoor Fireplaces, and Fixed Stationary Heating Units in Residential Structures

Data Source: 1993 - 1995 National Fire Incident Reporting System

26

Variations in the Rural Fire Problem — the North versus the South

In the course of investigating the rural fire problem in the U.S., separate analyses were conducted for the rural North versus the rural South and for the rural East versus the rural West. The East-West analysis resulted in few significant differences. The North-South analysis, in contrast, revealed several interesting facts, and these are discussed below. For subjects not addressed in this section, readers are referred to the earlier section on the characteristics of the fire problem in all rural areas. The topics addressed in this section include where fires occur, the causes of rural fires, extent of flame damage, area of fire origin, methods of fire extinguishment, smoke detector performance, and differences in heating fires.

Where Fires Occur

Similar to the U.S. as a whole, outside fires are the leading category of fires in both northern and southern rural areas, though the proportion is slightly lower in the North. For both the North and the South, structure fires are the second leading category of fires. However, as indicated in Table 4, the proportion of fires that occur in structures in rural areas of the North is somewhat higher than in the South.

**Table 4. Distribution of Northern and Southern Rural Fires
by Property Type**

Property Type	Percent Distribution	
	North	South
Outside Fires	43%	49%
Structure Fires	37%	32%
Non-Residential Structure Fires	12%	8%
Residential Structure Fires	25%	24%
Vehicle Fires	19%	18%
Other Fires	1%	1%
TOTAL FIRES	100%	100%

The average number of deaths that occur in each fatal fire is one measure for comparing how deadly different categories of fires are relative to one another. The average number of deaths per structure fire in rural areas of the North is higher than in rural areas of the South. In the North there are an average of 1.3 deaths per fatal fire.

The average in the South is lower at 1.2 (see Appendix D for a complete list of average numbers of deaths and injuries by type of fire).

The average number of injuries for all fires in which injuries were sustained provides another way of comparing fires. In northern rural areas, the average number of injuries resulting from structure fires is higher than in southern areas, but this difference pertains only to non-residential structures. Non-residential fires in northern rural areas have an average of 1.6 injuries per fire. In contrast, the average for southern rural areas is 1.4.

Causes of Fires

There are differences in causes of rural residential structure fires between the North and the South. Figure 22 displays the leading cause profiles for both regions. In the North, heating accounts for 39 percent of rural fires. Electrical fires are second (12 percent), cooking fires are third (11 percent), and arson fires are fourth (eight percent). In the South heating is the leading cause of fires, but it accounts for only 29 percent of rural fires, a lower proportion than in the North. The next three leading causes of rural fires in the South are cooking (18 percent), arson (12 percent), and electrical distribution (12 percent). In the rural South, cooking is a relatively greater cause of fires than in the North.

Area of Fire Origin

Figure 23 shows that residential structure fires originate most often in the chimney area in the rural North. The proportion of fires originating in chimneys in this region is three times higher than in the rural South. The second leading area of fire origin in the North is the kitchen or cooking area. The prevalence of fires that originate in chimneys in the rural North is expected due to the fact that heating fires occur relatively more frequently in that region.

In the rural South, residential structure fires originate most frequently in the kitchen or cooking area of the home, where they account for 21 percent of fires. The second most common area of origin in the rural South is sleeping rooms (12 percent).

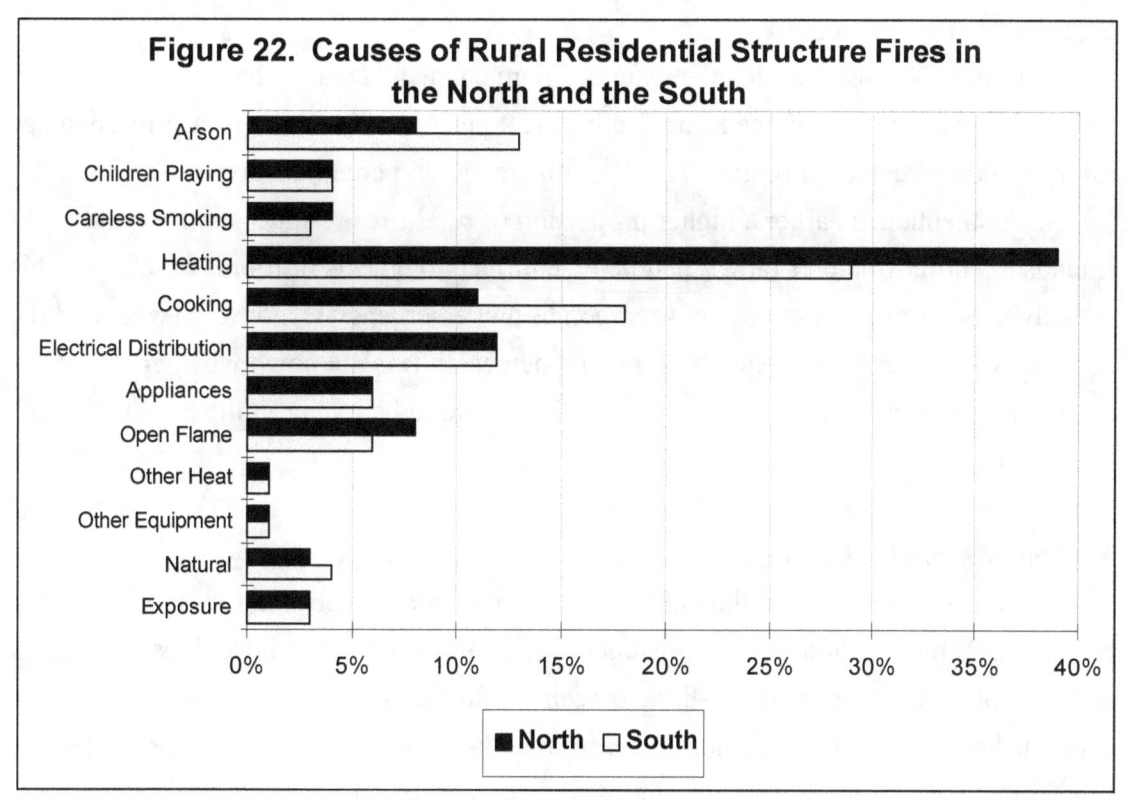

Figure 22. Causes of Rural Residential Structure Fires in the North and the South

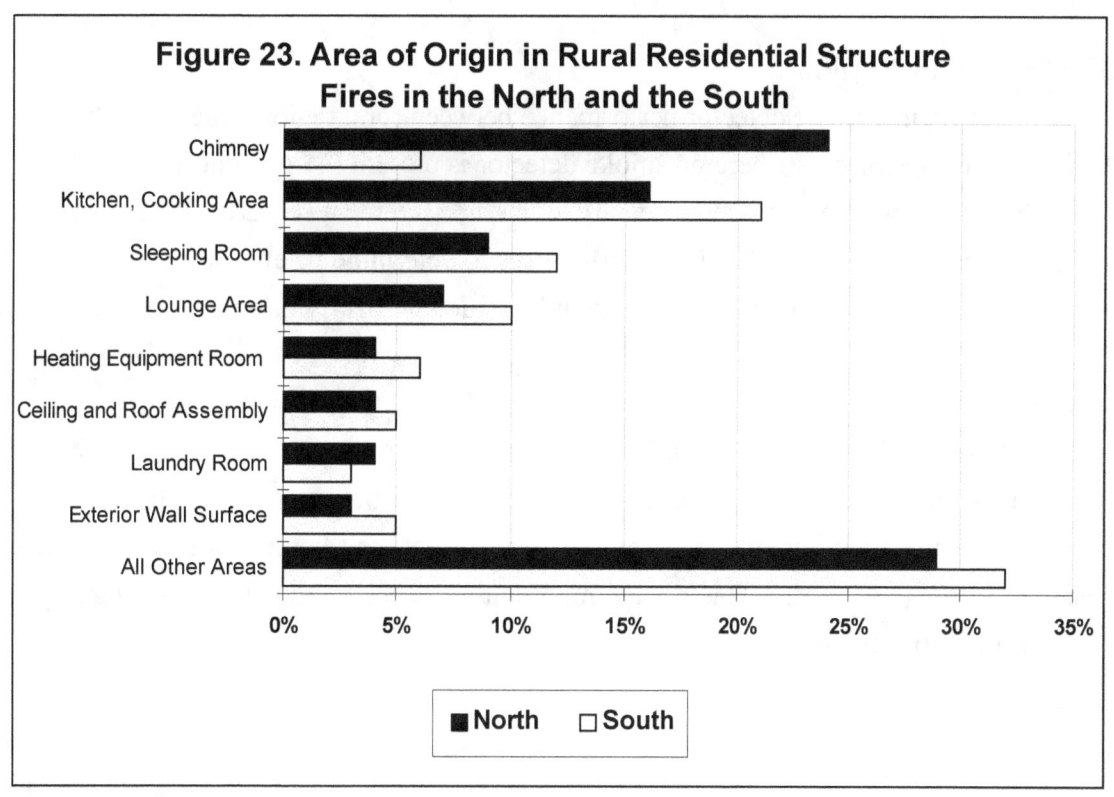

Figure 23. Area of Origin in Rural Residential Structure Fires in the North and the South

Data Source: 1993 - 1995 National Fire Incident Reporting System

Extent of Flame Damage

Figure 24 illustrates that the extent of flame damage tends to be more severe in residential structure fires in the rural South. In 34 percent of southern rural fires, damage extends to the structure of origin. This is compared to 26 percent of fires in the rural North. As mentioned earlier a higher proportion of rural fires result in damage that extends to entire structures rather than being confined to objects or rooms of origin. This is likely a factor of the distance between homes and emergency response services and the possibility that fires burn longer before being detected. It is unknown whether emergency response distances and response times tend to be longer in the South than in the North.

Method of Extinguishment

Figure 25 reveals that the only significant differences between the rural North and the rural South in method of extinguishment involve pre-connected hose lines from tanks and portable extinguishers. Fifty-three percent of fires in the rural South were extinguished by pre-connected hose lines from tanks. This compares to 42 percent of fires in the rural North. Eleven percent of fires in the rural North were extinguished by portable extinguishers compared to six percent in the rural South.

Smoke Detector Performance

Comparing smoke detector performance between rural regions, the South has a higher percentage of fires where no smoke detector is present (69 percent) (Figure 26). In the North this percentage is significantly lower, at 51 percent (Figure 27). Further research is needed to investigate these differences to determine whether the lack of smoke detectors is correlated with, for example, regional poverty or educational levels.

Heating Fires

As the leading cause of rural fires, it is important to have a clear understanding of the nature of heating fires. Because of climate differences between the North and the South, residents of rural areas in the South generally heat their homes for fewer days than residents of the rural North. This has an important influence on the profiles of heating fires for these two regions.

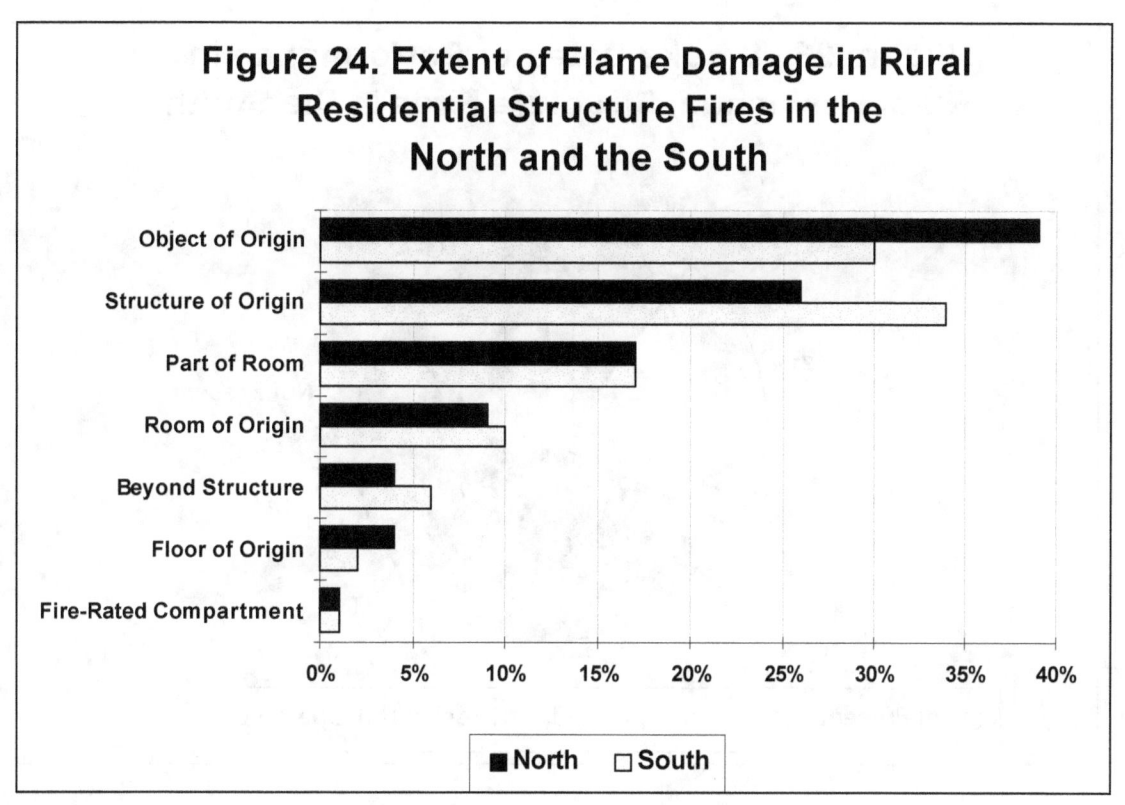

Figure 24. Extent of Flame Damage in Rural Residential Structure Fires in the North and the South

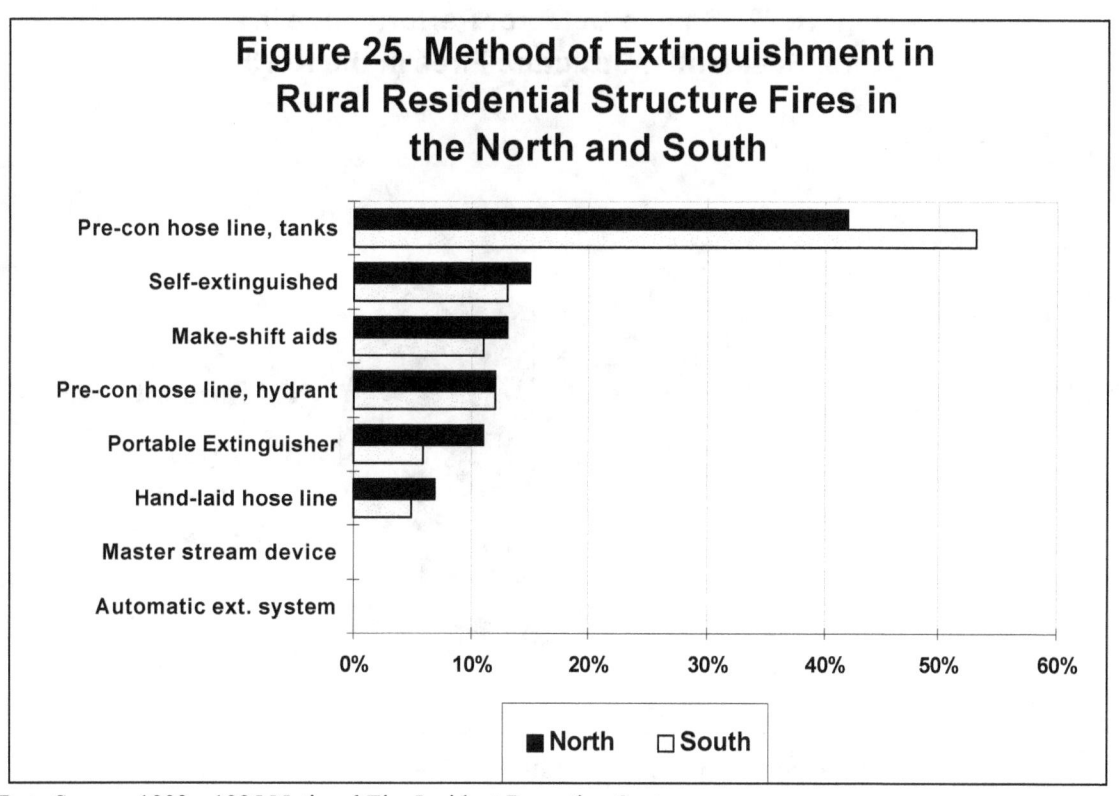

Figure 25. Method of Extinguishment in Rural Residential Structure Fires in the North and South

Data Source: 1993 - 1995 National Fire Incident Reporting System

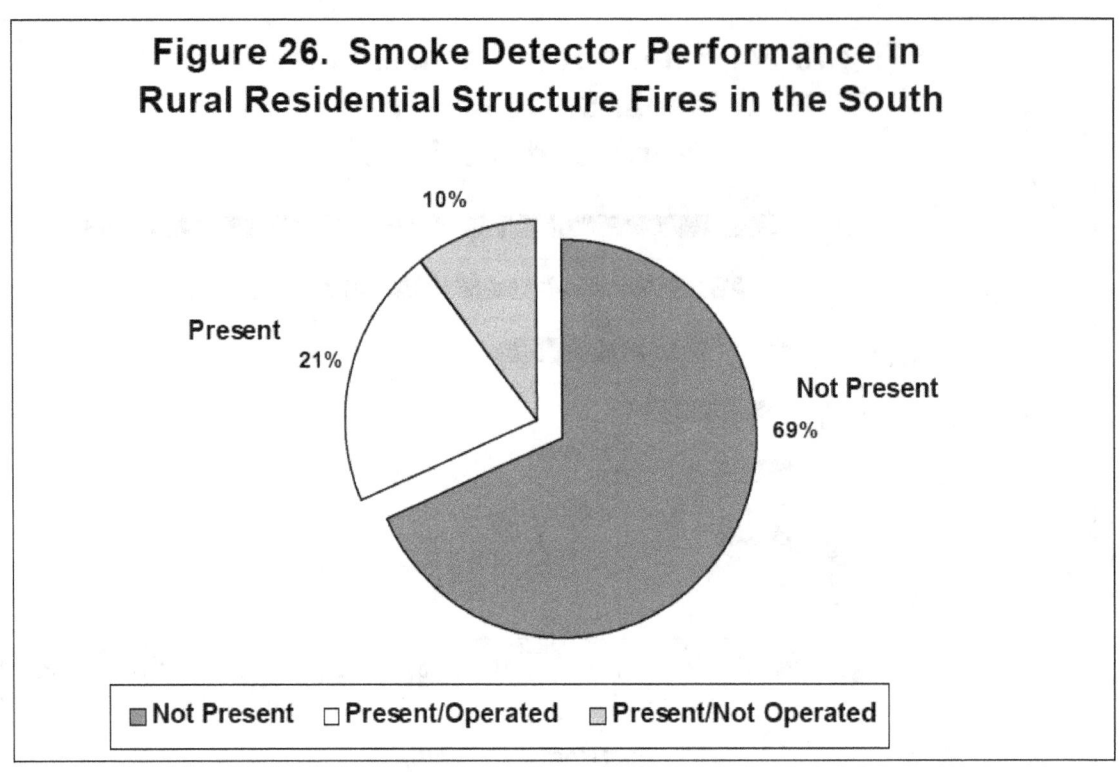

Figure 26. Smoke Detector Performance in Rural Residential Structure Fires in the South

10%

Present
21%

Not Present
69%

■ Not Present   □ Present/Operated   ▤ Present/Not Operated

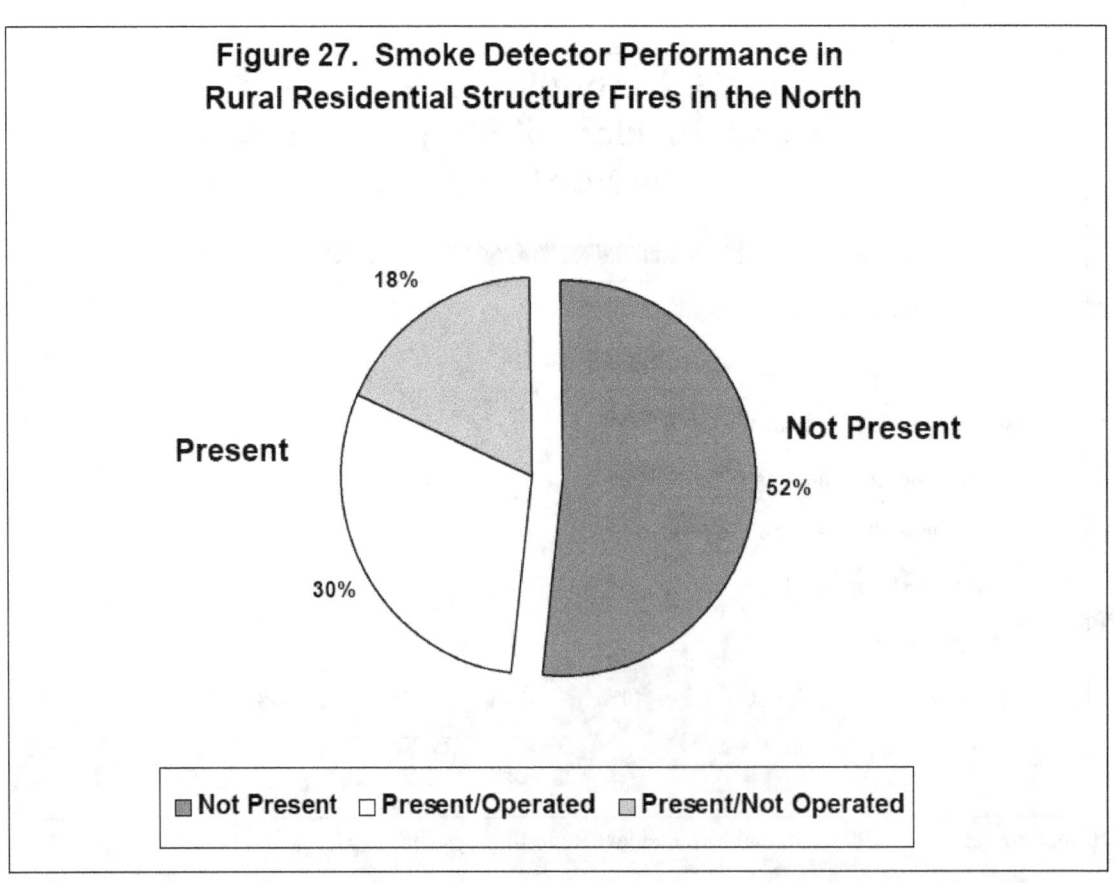

Figure 27. Smoke Detector Performance in Rural Residential Structure Fires in the North

18%

Present

Not Present
52%

30%

■ Not Present   □ Present/Operated   ▤ Present/Not Operated

Data Source: 1993 - 1995 National Fire Incident Reporting System

Equipment Involved in Ignition. While the leading pieces of equipment involved in ignition of rural heating fires are the same in the North and the South, their rank order is different. Figure 28 shows that in the North, fixed stationary heating units rank first (42 percent of heating fires) and chimneys rank second (23 percent). In the South, chimneys rank first (31 percent) and fixed stationary heating units rank second (27 percent). Because of climate differences, fewer southern homes may have fixed stationary heating units.

Type of Material Ignited. Figure 29 shows that among northern rural residential heating fires, the leading category of the type of material first ignited is adhesive, resin and tar, accounting for 53 percent of fires. Resin and tar are the compounds that build up in chimneys. The accumulation of resin and tar helps explain the high incidence of heating fires that originate in chimneys in the rural North. Adhesive, resin, and tar is also the leading category of the type of material first ignited (38 percent) in rural heating fires in the South. However the proportion of these fires is lower than in the rural North. In both areas, sawn wood is the second leading type of material ignited in rural heating fires.

Ignition Factor. Mechanical failure or malfunction is the overwhelming cause of rural heating fires in both the North and the South (Figure 30). As noted earlier, lack of maintenance or worn out equipment is responsible for a majority of heating fires due to mechanical failure or malfunction. In the North, 80 percent of these mechanical failure fires are due improperly maintained or worn out equipment. In the South, this percentage is 71 percent.

Fires that begin due to improperly maintained or worn out equipment suggest human error, at least indirectly. A much lower proportion of rural heating fires are due directly to operator deficiency. However, operator deficiency fires occur two times more often in heating fires in the rural North than in the rural South.

Manufactured Housing

Manufactured housing (often referred to as "mobile" or "trailer" homes) is a special category of residential dwelling. Although only a small fraction of the overall U.S. population lives in manufactured housing, in the past manufactured housing fires have represented a severe problem in terms of fire fatalities – they have had double the

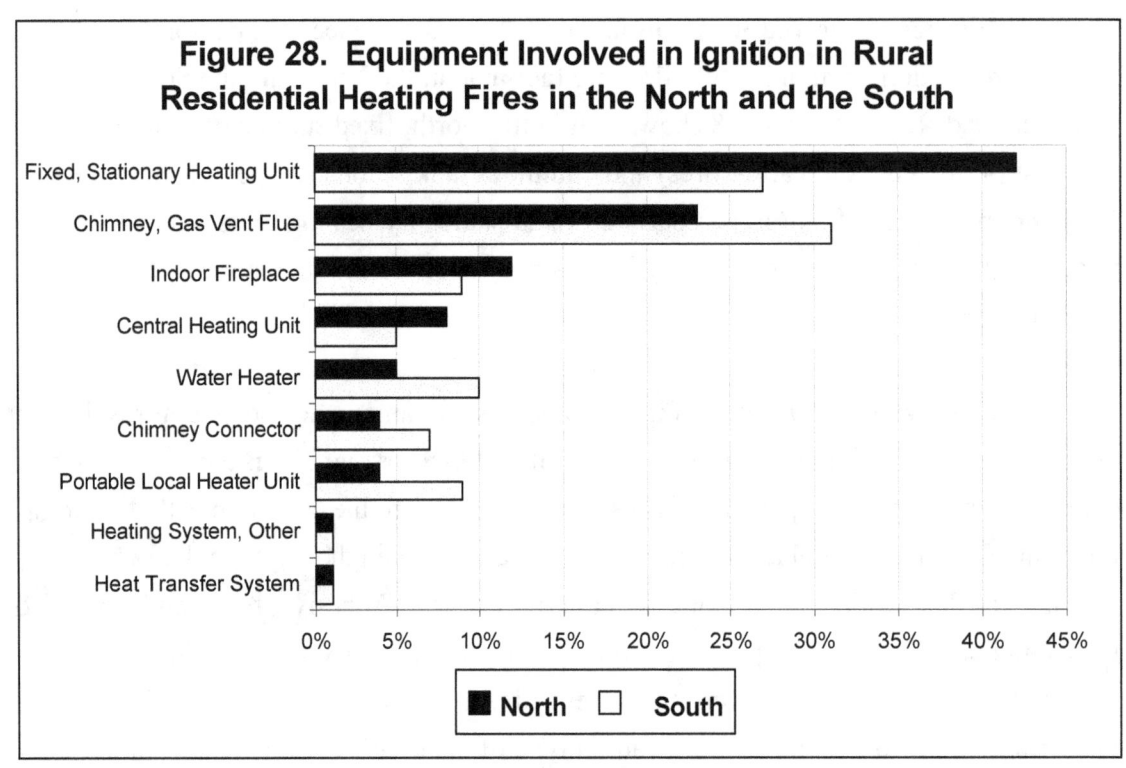

Figure 28. Equipment Involved in Ignition in Rural Residential Heating Fires in the North and the South

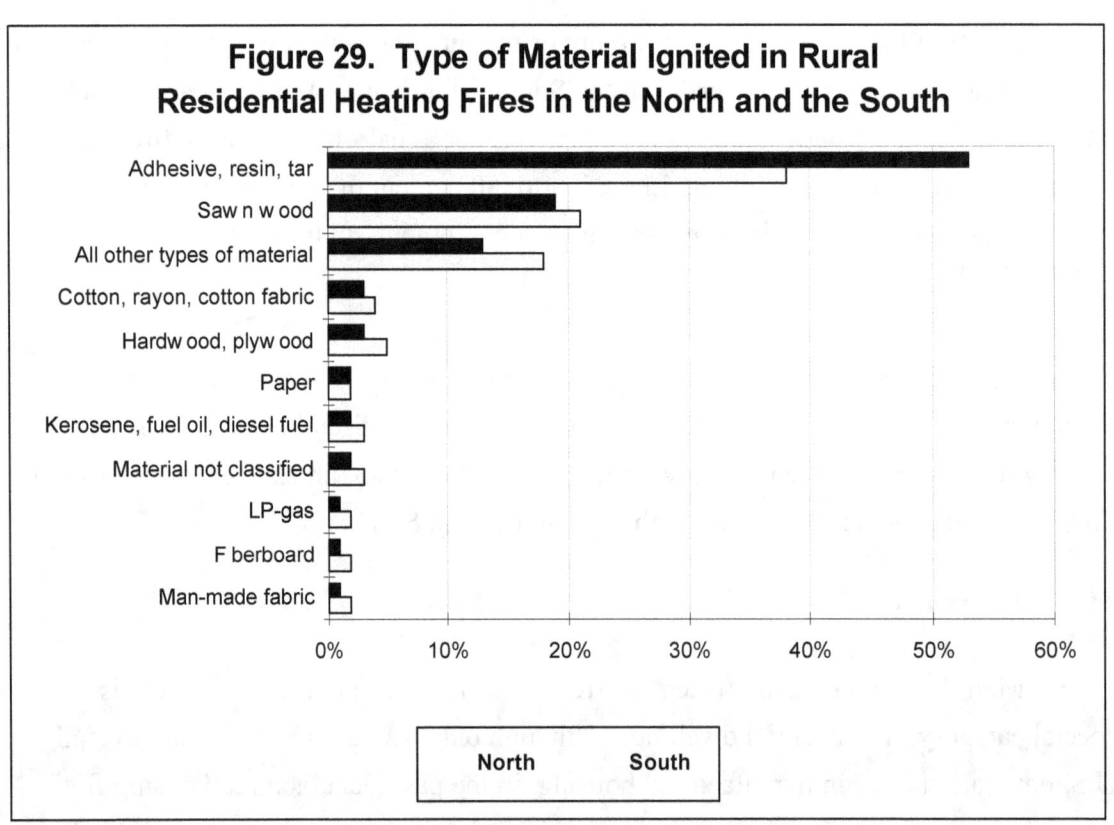

Figure 29. Type of Material Ignited in Rural Residential Heating Fires in the North and the South

Data Source: 1993 - 1995 National Fire Incident Reporting System

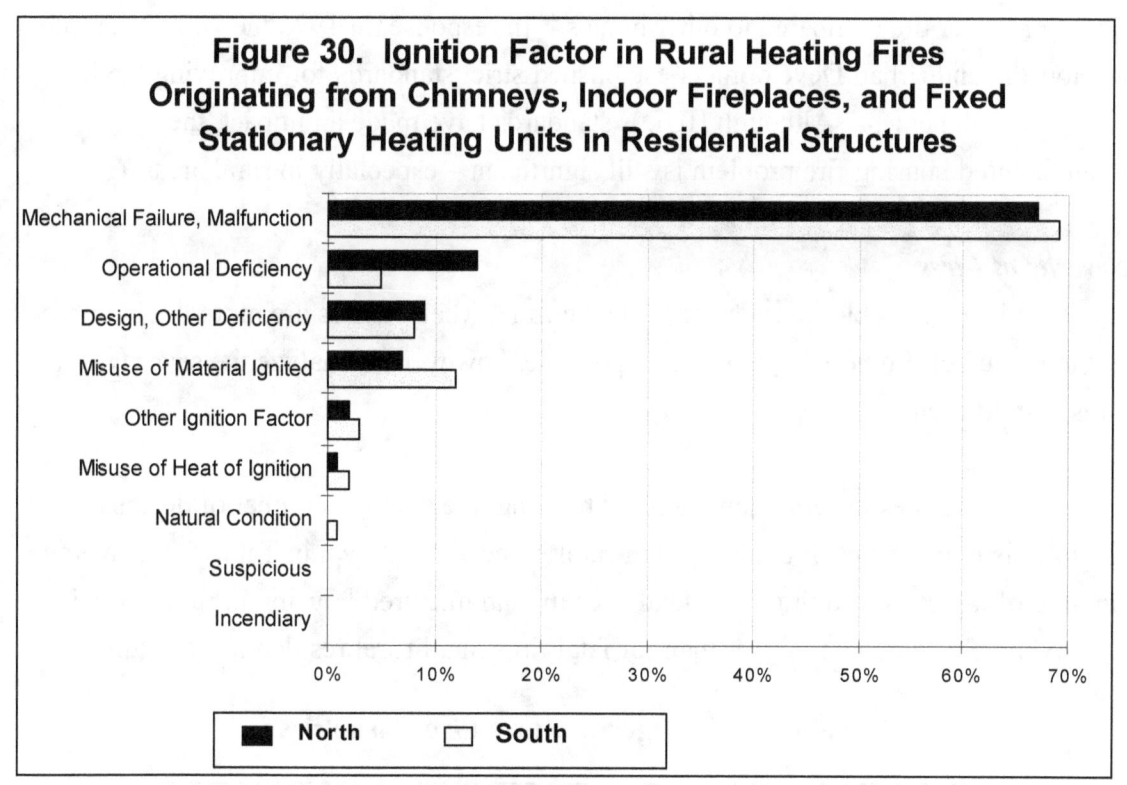

Figure 30. Ignition Factor in Rural Heating Fires Originating from Chimneys, Indoor Fireplaces, and Fixed Stationary Heating Units in Residential Structures

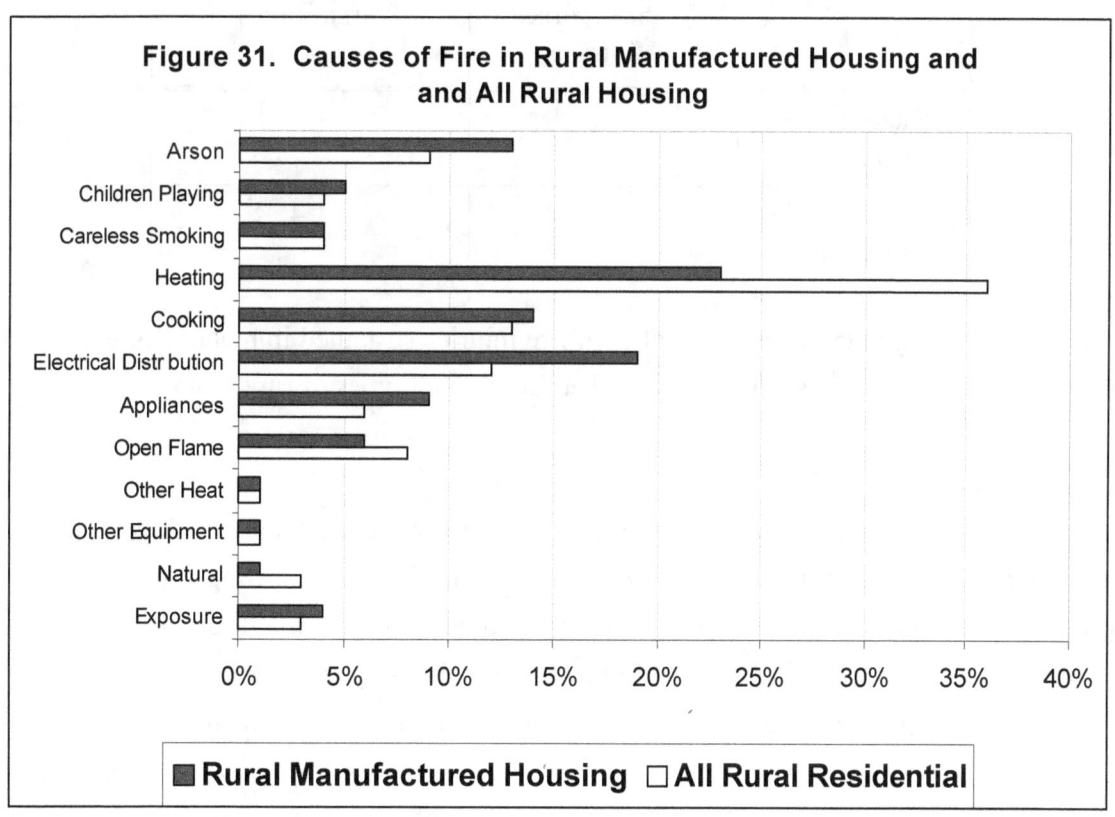

Figure 31. Causes of Fire in Rural Manufactured Housing and and All Rural Housing

Data Source: 1993 - 1995 National Fire Incident Reporting System

fatality rate per fire compared to other homes.[2] In response, in 1976 the U.S. Department of Housing and Urban Development established strict standards for improving the fire safety of such homes.[3] Although HUD's standards have made an impact, the manufactured housing fire problem is still significant – especially in rural areas.

Severity of Fires

It is not possible to use NFIRS to determine whether fires are more likely to break out in manufactured housing units. It is possible, however, to analyze the severity of the fires that do occur.

Among fires in rural manufactured housing, the average number of deaths per fatal fire is higher than for other types of rural homes. As shown in Table 5, the average number of fatalities resulting from fatal fires in manufactured housing in rural areas is approximately 15 percent higher than for fatal fires in all rural residential structures.

Table 5. Average Structure Fire Casualties

Casualty Type	Rural Manufactured Housing	All Rural Residential Structures
Fatalities per Fatal Fire	1.5	1.3
Injuries per Fire with Injuries	1.4	1.4

Interestingly, the average number of fire injuries sustained in manufactured housing fires with injuries is the same as that found in all types of rural homes.

Causes of Fires

Figure 31 indicates that the leading cause of manufactured housing fires in rural areas is heating, accounting for 23 percent of the total. However, heating is less predominate a cause of home fires in manufactured housing than in all rural residential structures. The other leading causes of manufactured housing fires include electrical distribution (19 percent) and cooking (14 percent). Comparing rural manufactured

---

[2] "Fire in the United States 1985-1994." Washington, DC: Federal Emergency Management Agency, United States Fire Administration, National Fire Data Center. Ninth Edition.
[3] Department of Housing and Urban Development Web site - Manufactured Housing Fact Sheet: http://www.hud.gov/fha/sfh/mhs/mhsshtmr html

housing to all other rural homes reveals that relatively more fires in manufactured housing are caused by electrical distribution, arson, and appliances.

Cooking fires are tied with heating fires as the leading cause of fires in manufactured housing in the rural South (19 percent). In contrast, heating fires are the leading cause and account for 26 percent of fires in manufactured housing in the rural North. Cooking fires account for only 11 percent of manufactured housing fires in the rural North. Again, this discrepancy is likely explained by climate: residents of the rural North are likely to heat their homes on more days than in the rural South, thus increasing the likelihood of experiencing a heating fire.

Smoke Detector Performance

As Figure 32 illustrates, a majority of rural manufactured housing fires occur in residences without smoke detectors (63 percent). If the number of manufactured housing fires without *operating* smoke detectors is included in this percentage, the proportion of homes without a functional smoke detector rises to 75 percent.

This problem is even more exaggerated in manufactured housing in the rural South. Seventy-three percent of these fires occur in homes without smoke detectors (Figure 33). Once non-operating detectors are considered, fully 81 percent of fires occur in structures that are not protected by a functional smoke detector. This compares to 70 percent of fires in the rural North (Figure 34).

The lack of operating smoke detectors in manufactured housing is of grave concern. These homes tend to be smaller than other types of homes, so a fire may more quickly engulf the entire structure, giving residents less time to escape than in other types of dwellings.

Extent of Flame Damage

Figure 35 shows that the most common levels of extent of flame damage in manufactured housing in rural areas are structure of origin, object of origin, and part of room, accounting for 47 percent, 20 percent, and 14 percent of all fires, respectively. The proportion of manufactured homes in which the extent of flame damage extends to the structure of origin is alarmingly high compared to the proportion of rural residential structures generally that sustain this amount of damage. In fact, manufactured housing in rural areas sustains flame damage that extends to their entire structures 62 percent more often than other rural residential structures. As discussed above, the size of these structures is an important factor to consider in understanding why the extent of flame damage is so much greater in these types of dwellings.

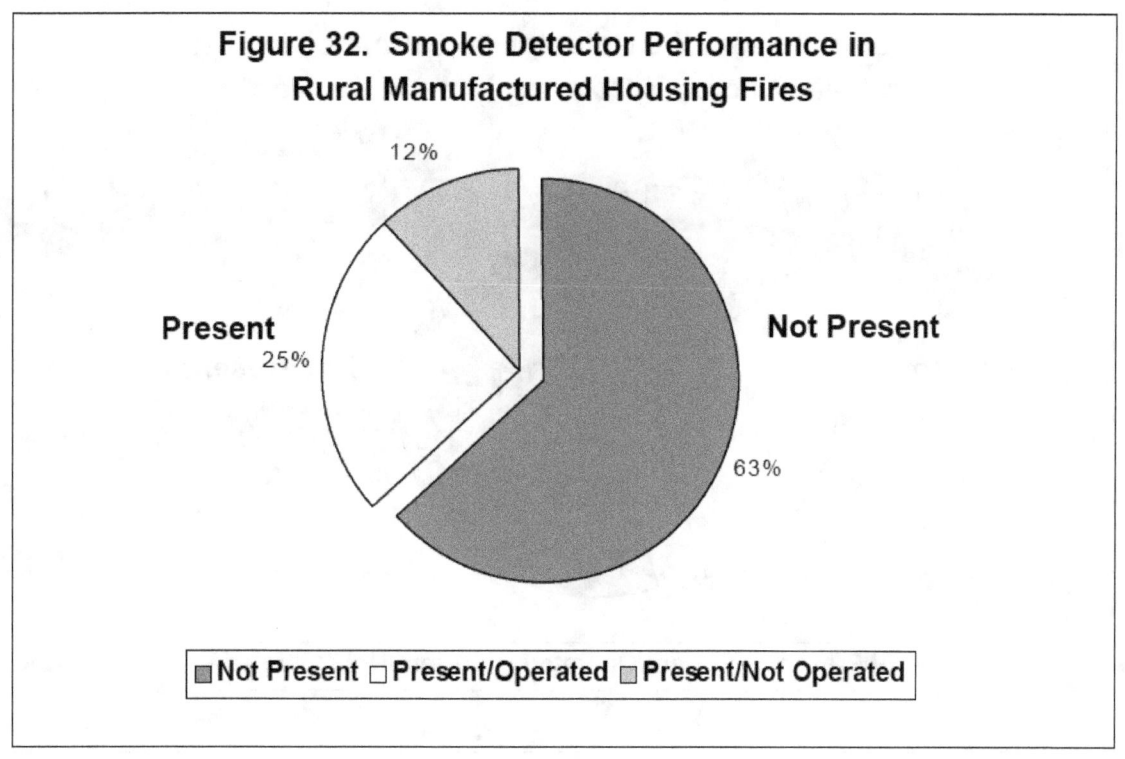

Figure 32. Smoke Detector Performance in Rural Manufactured Housing Fires

12%

Present
25%

Not Present

63%

■ Not Present □ Present/Operated ▨ Present/Not Operated

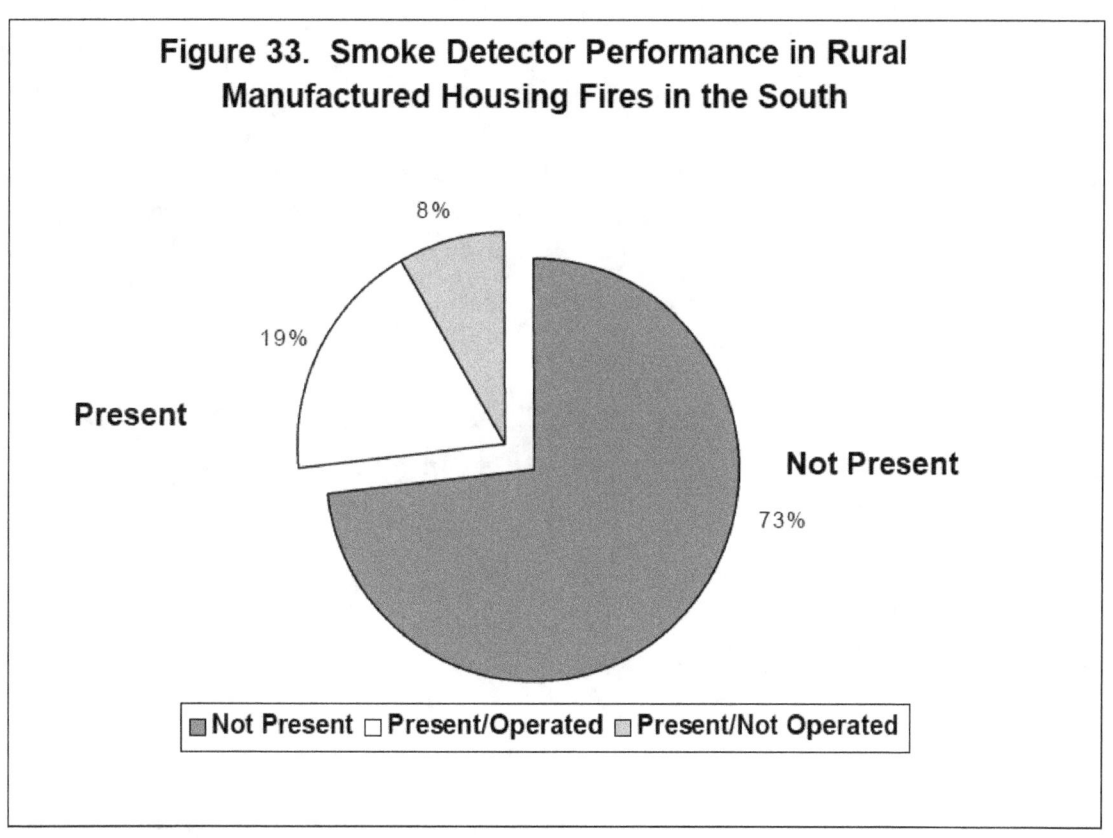

Figure 33. Smoke Detector Performance in Rural Manufactured Housing Fires in the South

8%

19%

Present

Not Present

73%

■ Not Present □ Present/Operated ▨ Present/Not Operated

Data Source: 1993 - 1995 National Fire Incident Reporting System

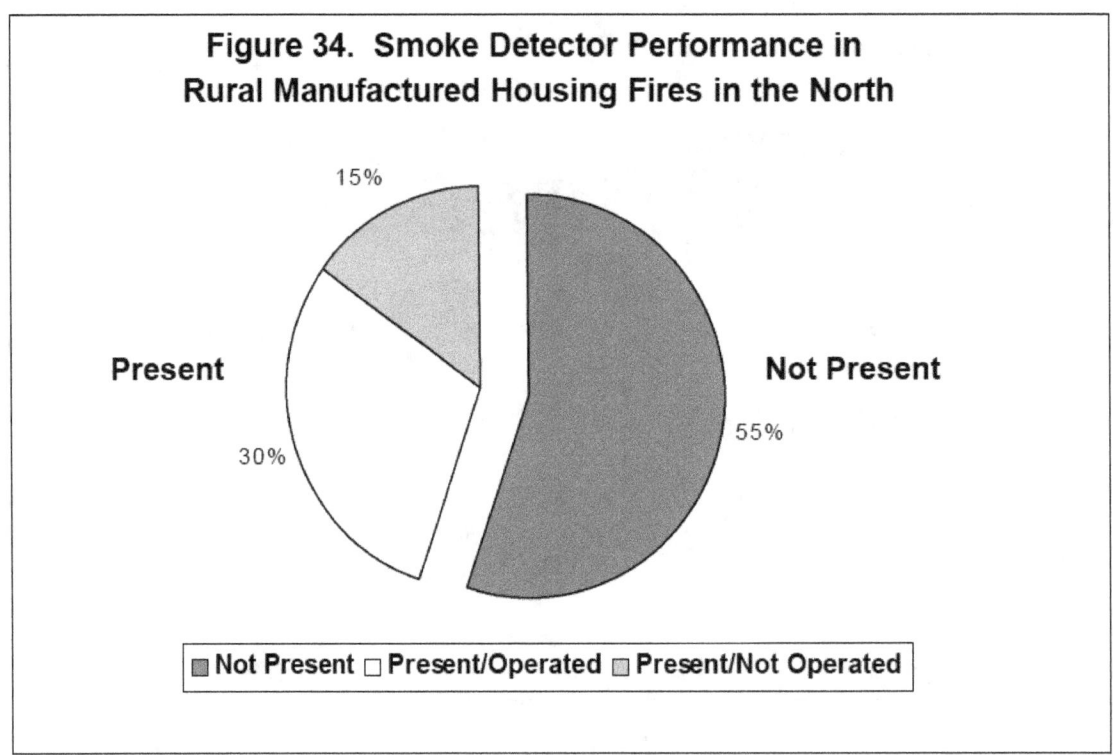

Figure 34. Smoke Detector Performance in Rural Manufactured Housing Fires in the North

Present

Not Present

15%

30%

55%

■ Not Present □ Present/Operated ▨ Present/Not Operated

Data Source: 1993 - 1995 National Fire Incident Reporting System

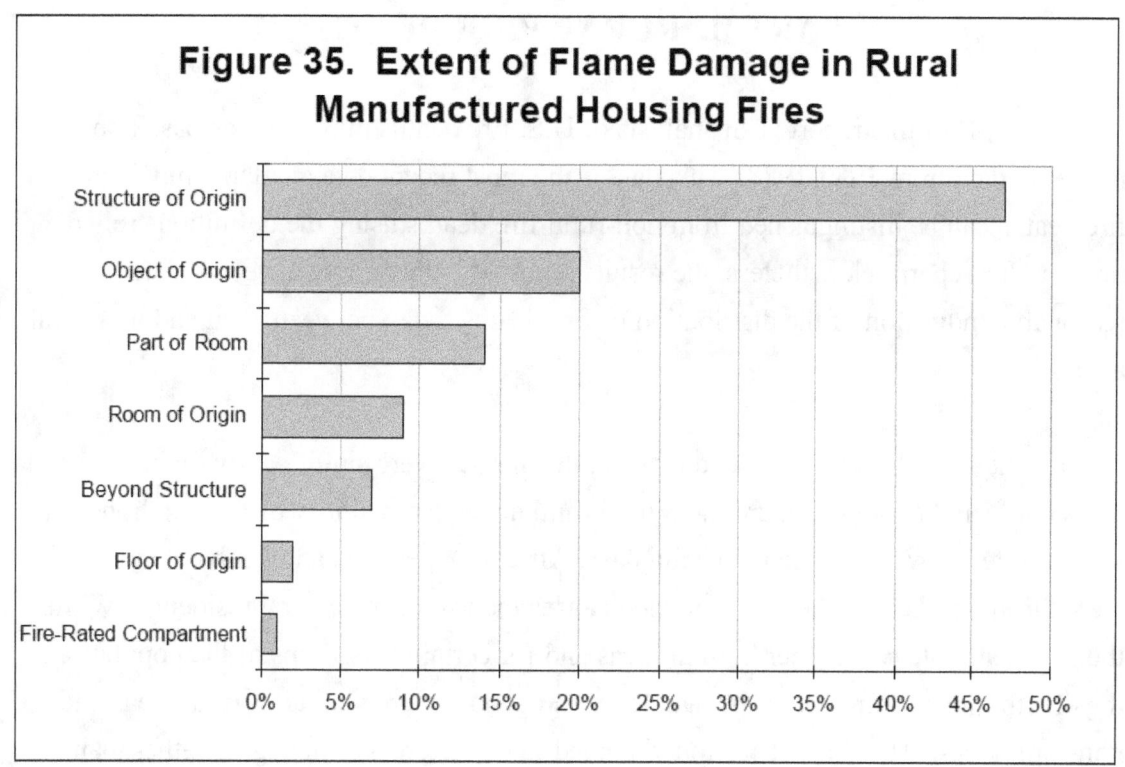

Figure 35. Extent of Flame Damage in Rural Manufactured Housing Fires

Data Source: 1993 - 1995 National Fire Incident Reporting System

PART II. RURAL FIRE DEATHS

Part II of this report is an analysis of U.S. fire deaths in rural areas based on mortality data from 1983-1988. This data is the most recent data available in which rural fire deaths can be distinguished from non-rural fire deaths using the definitions relied upon in this report. Nonetheless, the results reported below are believed to be a reasonable indication of the distribution of fire deaths today between rural and non-rural areas.

The mortality data revealed that fire death rates were significantly higher in rural areas compared to non-rural areas. Within rural areas, the majority of fire death victims were White. However, on a per capita basis African Americans and Native Americans were far more likely to die of fire-related causes than were other rural residents. While the fire death rate was higher in rural areas and for certain subgroups of the population, the distributions of fire deaths by age, race, and gender were similar in rural areas and non-rural areas. The differences found tended to be differences in degree rather than suggesting different underlying patterns altogether.

Fire Death Rates in Rural Areas

Table 6 contains fire death rates for rural and non-rural areas. The fire death rates are presented for the population as a whole and by ethnicity. Table 6 shows that fire death rates over the 1983-1988 time period were 36 percent higher in rural areas than non-rural areas. Whereas the fire death rate in non-rural areas was 22.8 per million population, the rate in rural areas was 30.9. These differences became even greater when fire death rates were broken down by race and ethnicity.

Table 6. U.S. Fire Deaths Rates in Rural and Non-Rural Areas

Rural/Non-rural	1993 Population	1983-1988 Fire Deaths	Average Annual Fire Deaths	Fire Deaths per Million Population
Total U.S. Population	**245,139,132**	**34,584**	**5,764**	**23.5**
Rural	21,840,693	4,054	676	30.9
Non-rural	223,298,439	30,530	5,088	22.8
White Population	**205,834,339**	**23,681**	**3,947**	**19.2**
Rural	19,382,873	2,877	480	24.7
Non-rural	186,451,466	20,804	3,467	18.6
African American Population	**29,621,848**	**10,213**	**1,703**	**54.8**
Rural	1,879,767	999	167	88.6
Non-rural	27,742,081	9,214	1,536	55.4
Native American Population	**2,096,447**	**398**	**66**	**31.5**
Rural	461,521	168	28	60.7
Non-rural	1,634,926	230	38	23.4
Asian and Pacific Islander Pop.	**7,586,498**	**132**	**22**	**2.9**
Rural	116,532	6	1	8.6
Non-rural	7,469,966	126	21	2.8

Table 6 shows that the overall fire death rate for African Americans was over two times that of the general U.S. population. This disparity was even greater in rural areas. Rural African Americans had a death rate of 88.6 per million population. This was over three and one-half times greater than the fire death risk for rural Whites. Similarly, rural Native Americans had a fire death rate of 60.7 per million, a rate that was almost two and one-half times greater than that of rural Whites.

But while rural African and Native Americans suffered significantly higher fire death risks than other groups, the absolute numbers of their fire deaths were relatively more modest, especially for Native Americans. Whereas an average of 480 fire death victims each year were White, an average of 167 fire death victims were African American and 28 were Native American.

The Distribution of Fire Deaths in Rural Areas

Fire Deaths by Race

As Figure 36 indicates, the majority of rural fire deaths were White, with African Americans second, Native Americans third, and Asian and Pacific Islanders last. These numbers are indicative of the relative populations of these ethnic groups in rural areas.

Fire Deaths by Gender

Figure 37 shows that the distribution of male and female fire deaths was similar in both rural and non-rural populations. Nearly twice as many men died in fires as women.

Fire Deaths by Age Group

Overall, rural fire deaths tended to affect a larger portion of the younger population than non-rural fire deaths (Figure 38). This was a direct result of the dominance of the White fire death trend.

Fire Deaths among Whites by Age Group

As Figure 39 indicates the rural and non-rural age profiles for white fire deaths were very similar, but the age distribution was skewed slightly towards the younger end for the rural population. The very young, the very old, and middle-aged population groups shared the same percentage of fire deaths. Rural young people aged 1-24 were represented in higher proportions than their non-rural counterparts. Accordingly, the older population (55 to 75) accounted for a slightly lower proportion of fire deaths. Young adults aged 25-34 represented the largest single group for both rural and non-rural groups.

Fire Deaths among African Americans by Age Group

Like whites, the rural and non-rural African American age profiles were very similar. What was different is that, unlike the white population, the profile was slightly skewed to the older end for the rural population. A larger percentage of very elderly (over 75 years) rural African Americans died in fires than their non-rural counterparts and, accordingly, a smaller percentage of rural middle-aged and older African American died in fires (Figure 40).

Fire Deaths among Native Americans by Age Group

Rural fire deaths for Native Americans were highly skewed to the youngest ages. The very young, those under five years of age, represented almost over 30 percent of all Native American fire deaths (Figure 41).

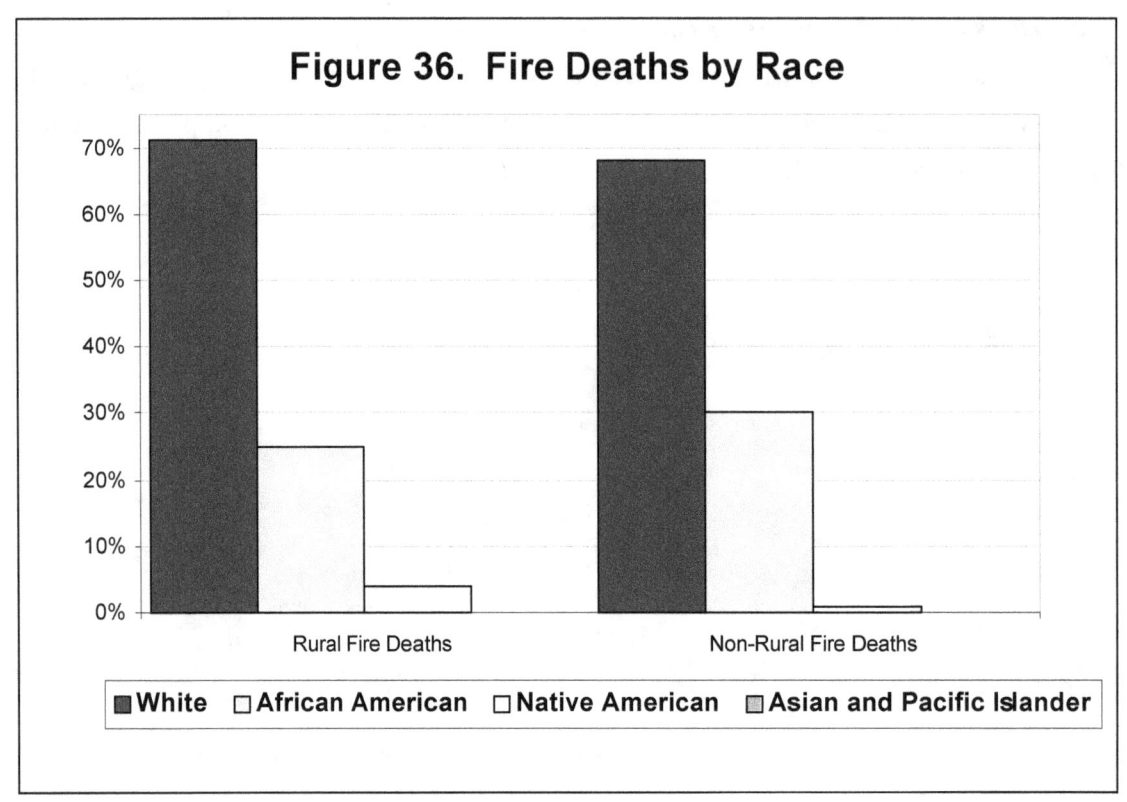

Figure 36. Fire Deaths by Race

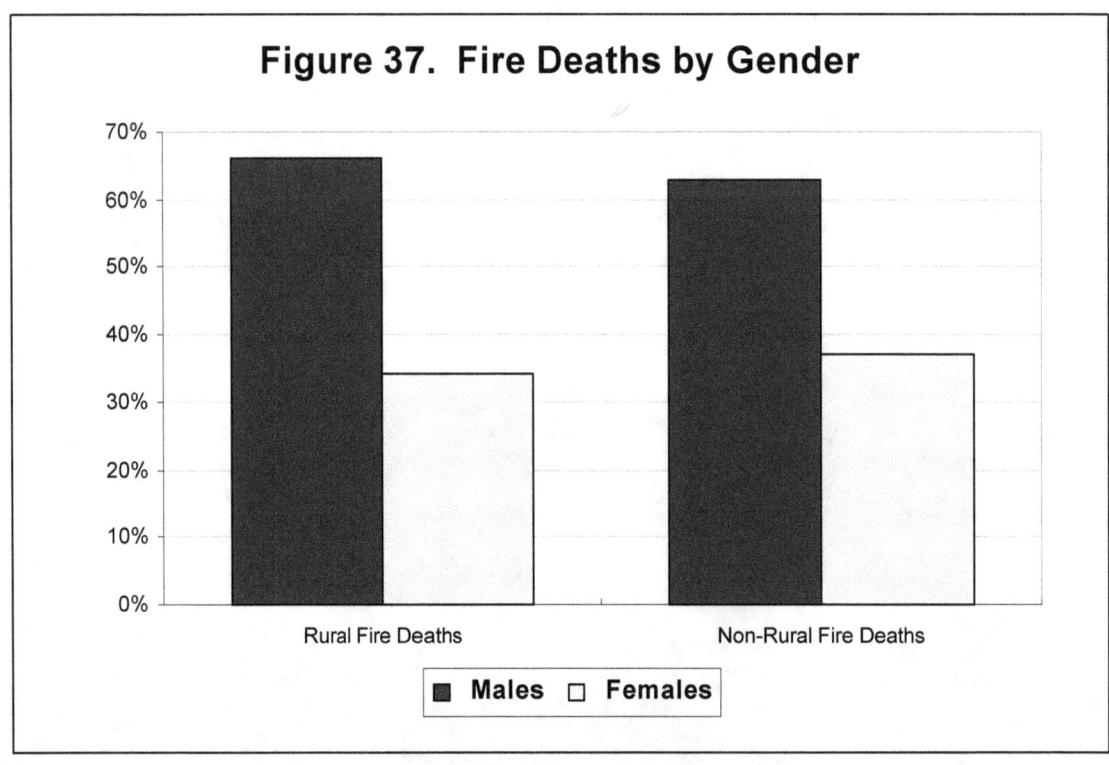

Figure 37. Fire Deaths by Gender

Data Source: 1993 - 1995 National Fire Incident Reporting System

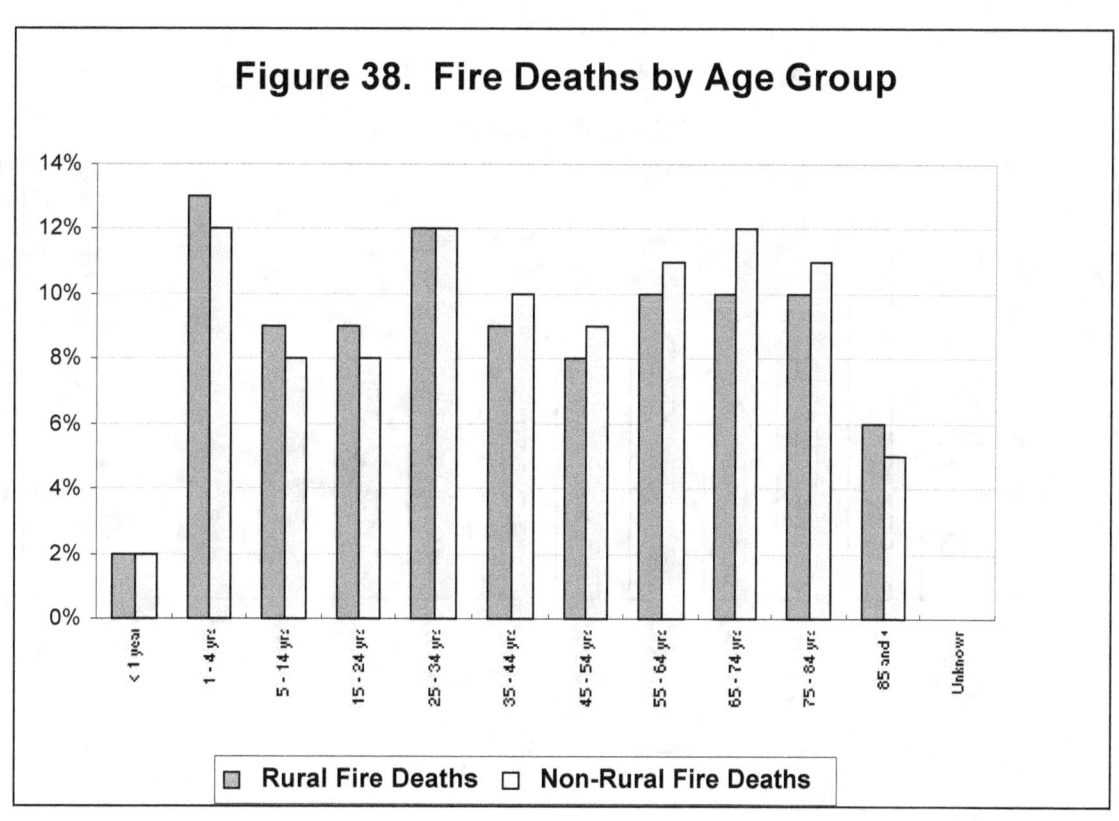

Figure 38. Fire Deaths by Age Group

■ Rural Fire Deaths □ Non-Rural Fire Deaths

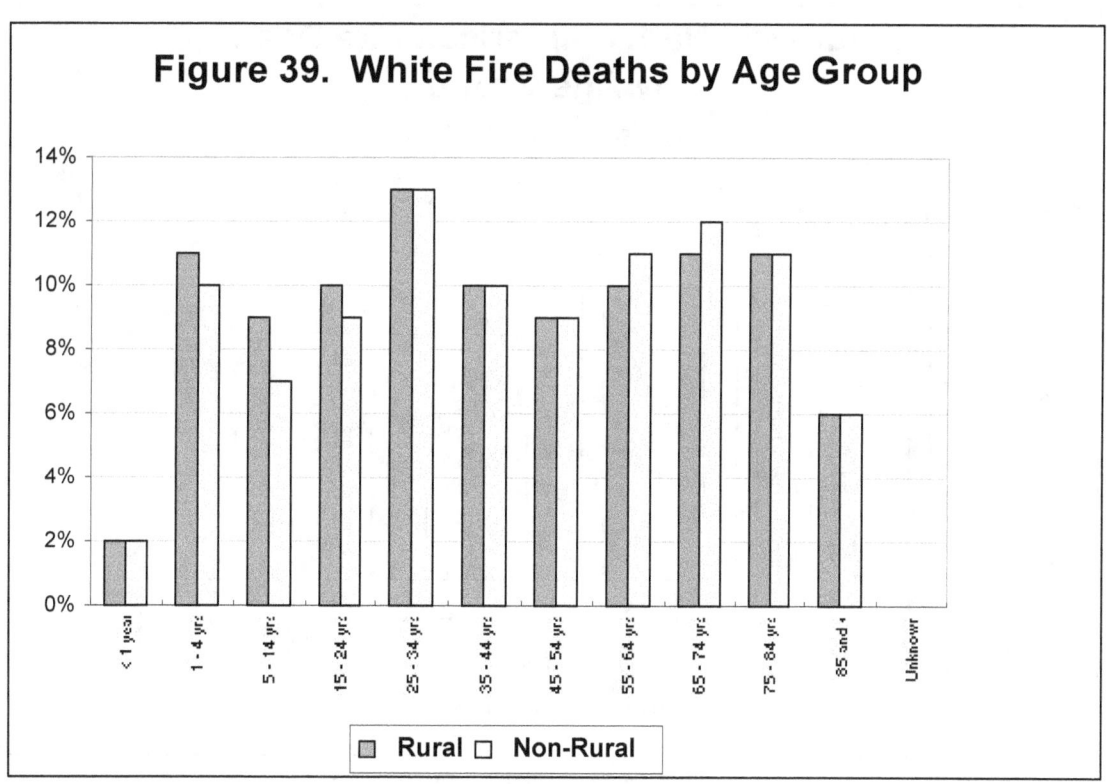

Figure 39. White Fire Deaths by Age Group

■ Rural □ Non-Rural

Data Source: 1993 - 1995 National Fire Incident Reporting System

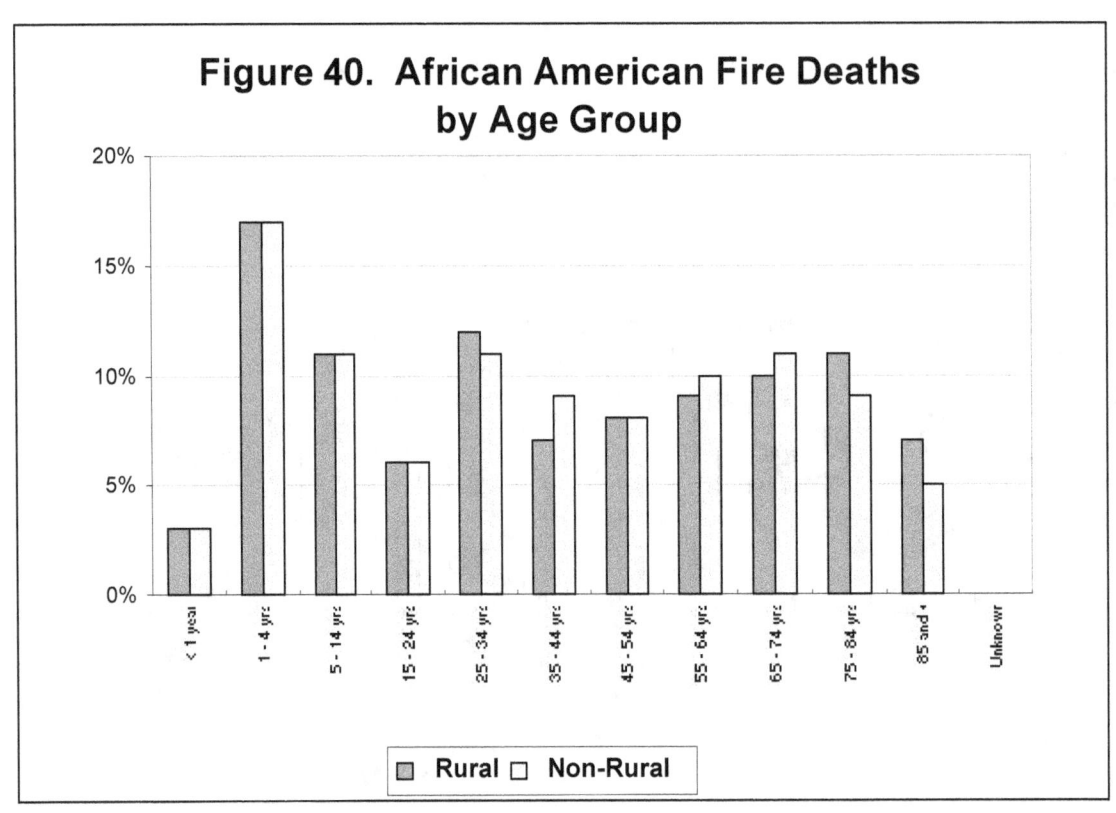

Figure 40. African American Fire Deaths by Age Group

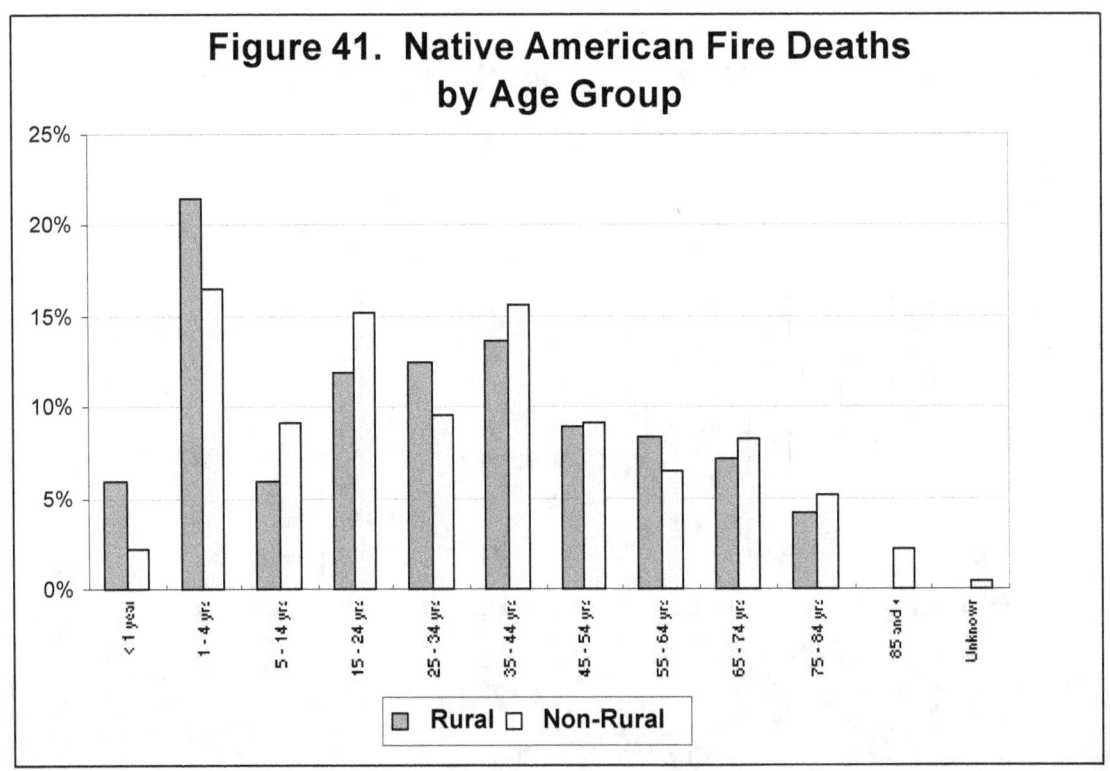

Figure 41. Native American Fire Deaths by Age Group

Data Source: 1993 - 1995 National Fire Incident Reporting System

APPENDICES A-D

APPENDIX A. - NFIRS GLOSSARY OF TERMS

Term	Definition
Area of Origin	The area of origin may be a room, an area or portion of a room, a vehicle or possibly some open area devoted to a specific use.
Construction Type	The type of building construction used in the structure where the fire occurred.
a - Fire resistive	A totally non-combustible building in which no structural steel is exposed and all vertical openings are protected with approved doors. The fire resistant coverings of the steel is typically very heavy; poured concrete , brick, concrete block, or similar material.
b - Heavy timber	A typical mill-constructed building in which the load-bearings walls or columns are masonry or heavy timber and all exposed wood members have a minimum dimension of two (2) inches. If steel or irons columns are used, they should be protected by a fire-resistance enclosure.
c - Protected Non-combustible	A totally non-combustible building in which no structural steel is exposed. All vertical openings are protected by approved doors. The fire-resistant covering of the steel is typically light: gypsum board, sprayed fire resistive covering, rated ceilings, and similar materials.
d - Unprotected Non-combustible	A totally non-combustible building in which the structural steel is exposed to the effects of a fire.
e - Protected Ordinary	The load-bearing walls are masonry. Columns are protected by a fire-resistive covering. The underside of all wood floor and roof decks is protected by a fire-resistive coating.
f - Unprotected Ordinary	The load-bearing walls are masonry. Columns, wood floor and roof decks are exposed and unprotected from fire.
g - Protected Wood Frame	Walls, floors and roof structure are wood framing. The interior wall and ceiling surfaces of a habitable spaces are protected by a fire resistive covering. A brick veneer building falls in this category because the wall structure is wood framed. But for any wood frame building if the basement does not have a fire-resistive ceiling protecting the underside of the first floor, the building should be classified in the "unprotected wood frame" category.

Term	Definition
h - Unprotected Wood Frame	Walls, floors, and roof structure are wood framing. There is no fire-resistive covering protecting the wood frame. A typical residential garage would fall in this category.
Equipment Involved in Ignition	The piece of equipment, if any, which provided the principal heat that caused ignition, whether the equipment malfunctioned or was used improperly.
Extent of Flame Damage	The size of the fire in terms of how far the flame damage extended. The extent of flame damage is the area that was actually burned or charred and not the area that received only heat, smoke, or water damage.
Ignition Factor	The condition or situation that allowed the heat source and combustible material to combine to start a fire. For example, the ignition factor can be a deliberate act, a mechanical failure, or an act of nature.
Manufactured Housing	This is a special category of one- and two- family dwellings. Includes private dwellings and duplexes with total sleeping accommodations for no more than 20 persons. Also includes mobile homes provided they are not in transit. It does not include townhouses or rowhouses.
Method of Extinguishment	This data element identifies how the fire was extinguished. This includes actions taken by the fire department, other people in the area, and automatic sprinkler or other fire extinguishing systems.
Non-residential Structure Fires	All fires which occur on property including industrial and commercial properties, institutions (such as hospitals, nursing homes, prisons), educational establishments, mobile properties, and properties that are vacant or under construction.
Other Fires	All fires which don't fit under the categories of outdoor, structure, or vehicle. Includes explosions without fires or accompanying fires.
Outdoor Fires	All fires outside of structures other than vehicle fires. Includes yard storage, crops or any fire outside a structure where the material burning has value.
Residential Structure Fires	All fires which occur on property including one- or two-family dwellings or multifamily apartment buildings. It also includes manufactured housing, hotels and motels, residential hotels, dormitories, and halfway houses.

Term	Definition
Smoke Detector Performance	The existence and location of fire detection equipment relative to the area of fire origin and whether the detection equipment worked.
Structure Fires	Any fire inside a structure, or on, under, or touching a structure.
Type of Material Ignited	The composition of the material which was first ignited by the heat source. This term refers to the raw, common, or natural state in which the material exists.
Vehicle Fires	All fires centered around means of transportation.

Source: National Fire Incident Reporting System Handbook Version 4.1. Washington, DC: Federal Emergency Management Agency, United States Fire Administration. 1989.

Appendix B. North/South and East/West Coding System

State	State Abbreviation	State FIPS Code	North/South Code	East/West Code
Alabama	AL	1	1	0
Alaska	AK	2	0	1
Arizona	AR	4	1	1
Arkansas	AR	5	1	1
California	CA	6	1	1
Colorado	CO	8	0	1
Connecticut	CT	9	0	0
Delaware**	DE	10	0	0
District of Columbia	DC	11	1	0
Florida	FL	12	1	0
Georgia	GA	13	1	0
Hawaii**	HI	15	1	1
Idaho	ID	16	0	1
Illinois	IL	17	0	0
Indiana**	IN	18	0	0
Iowa	IA	19	0	1
Kansas	KS	20	0	1
Kentucky	KY	21	1	0
Louisiana	LA	22	1	1
Maine**	ME	23	0	0
Maryland	MD	24	1	0
Massachusetts	MA	25	0	0
Michigan	MI	26	0	0
Minnesota	MN	27	0	1
Mississippi**	MS	28	1	0
Missouri	MO	29	0	1
Montana	MT	30	0	1
Nebraska	NE	31	0	1
Nevada**	NV	32	1	1
New Hampshire	NH	33	0	0
New Jersey	NJ	34	0	0
New Mexico**	NM	35	1	1
New York	NY	36	0	0
North Carolina**	NC	37	1	0
North Dakota**	ND	38	0	1
Ohio	OH	39	0	0
Oklahoma	OK	40	1	1
Oregon	OR	41	0	1
Pennsylvania**	PA	42	0	0
Rhode Island	RI	44	0	0
South Carolina	SC	45	1	0
South Dakota	SD	46	0	1
Tennessee	TN	47	1	0
Texas	TX	48	1	1
Utah	UT	49	0	1
Vermont	VT	50	0	0
Virginia	VA	51	1	0
Washington	WA	53	0	1
West Virginia	WV	54	0	0
Wisconsin	WI	55	0	0
Wyoming	WY	56	0	1

KEY:
North/South:
0 = North
1 = South
East/West:
0 = East
1 = West

** Indicates a state not participating in NFIRS

Appendix C. Average Number of Fires, Deaths, and Injuries by Major Categories of Fire, 1993-1995

By Property Type	ALL AREAS*						NORTHERN AREAS						SOUTHERN AREAS					
	ALL*		NON-RURAL		RURAL		ALL		NON-RURAL		RURAL		ALL		NON-RURAL		RURAL	
Number of Fires																		
Total	878,942		767,364		61,416		464,863		413,925		38,555		414,077		353,438		22,858	
Outside	381,487	43%	330,498	43%	27,624	45%	188,102	40%	166,379	40%	16,417	43%	193,383	47%	164,118	46%	11,205	49%
Structure	275,279	31%	240,835	31%	21,486	35%	158,396	34%	140,615	34%	14,140	37%	116,883	28%	100,220	28%	7,346	32%
Non-residential Structures	77,278	9%	67,124	9%	6,304	10%	49,024	11%	43,446	10%	4,521	12%	28,254	7%	23,678	7%	1,783	8%
Residential Structures	198,001	23%	173,711	23%	15,182	25%	109,372	24%	97,169	23%	9,619	25%	88,629	21%	76,542	22%	5,563	24%
Vehicle	207,941	24%	184,110	24%	11,509	19%	111,603	24%	101,333	24%	7,491	19%	96,338	23%	82,777	23%	4,017	18%
Other	14,235	2%	11,921	2%	797	1%	6,762	1%	5,598	1%	507	1%	7,473	2%	6,323	2%	290	1%
Number of Deaths																		
Total	2,232		1,871		247		1,231		1,052		170		1,002		818		75	
Outside	58	3%	38	2%	9	4%	27	2%	20	2%	7	4%	32	3%	18	2%	3	4%
Structure	1,737	78%	1,483	79%	181	73%	998	81%	866	82%	125	74%	739	74%	617	75%	56	75%
Non-residential Structures	125	6%	107	6%	10	4%	89	7%	79	8%	8	5%	37	4%	27	3%	2	3%
Residential Structures	1,612	72%	1,376	74%	171	69%	909	74%	787	75%	117	69%	702	70%	590	72%	54	72%
Vehicle	375	17%	297	16%	51	21%	174	14%	139	13%	33	19%	201	20%	158	19%	15	20%
Other	62	3%	53	3%	6	2%	32	3%	27	3%	5	3%	30	3%	25	3%	1	1%
Number of Injuries																		
Total	15,128		13,634		700		8,619		8,044		466		6,509		5,590		236	
Outside	741	5%	612	4%	40	6%	331	4%	299	4%	27	6%	410	6%	313	6%	14	6%
Structure	12,256	81%	11,291	83%	530	76%	7,304	85%	6,859	85%	352	76%	4,952	76%	4,432	79%	178	75%
Non-residential Structures	1,962	13%	1,788	13%	107	15%	1,310	15%	1,216	15%	81	17%	652	10%	571	10%	26	11%
Residential Structures	10,294	68%	9,503	70%	423	60%	5,994	70%	5,643	70%	271	58%	4,300	66%	3,861	69%	152	64%
Vehicle	1,518	10%	1,186	9%	93	13%	631	7%	564	7%	61	13%	887	14%	622	11%	32	14%
Other	613	4%	545	4%	37	5%	353	4%	322	4%	26	6%	260	4%	223	4%	12	5%
Fatal Fires																		
Total	1,738		1,446		198		953		813		129		784		634		66	
Outside	51	3%	32	2%	6	3%	26	3%	22	3%	2	2%	25	3%	12	2%	2	3%
Structure	1,345	77%	1,143	79%	144	73%	755	79%	653	80%	96	74%	589	75%	489	77%	48	73%
Non-residential Structures	89	5%	75	5%	8	4%	60	6%	53	7%	6	5%	29	4%	22	3%	2	3%
Residential Structures	1,256	72%	1,068	74%	136	69%	695	73%	600	74%	90	70%	560	71%	467	74%	46	70%
Vehicle	294	17%	231	16%	42	21%	145	15%	116	14%	27	21%	149	19%	115	18%	15	23%
Other	48	3%	40	3%	6	3%	27	3%	22	3%	2	2%	21	3%	18	3%	1	2%
Number of Fires with Injuries																		
Total	10,935		9,953		516		6,108		5,678		340		4,827		4,275		176	
Outside	603	6%	514	5%	34	7%	280	5%	252	4%	23	7%	323	7%	262	6%	11	6%
Structure	8,757	80%	8,058	81%	382	74%	5,053	83%	4,727	83%	251	74%	3,704	77%	3,331	78%	131	74%
Non-residential Structures	1,258	12%	1,139	11%	71	14%	829	14%	765	13%	52	15%	427	9%	373	9%	18	10%
Residential Structures	7,499	69%	6,919	70%	311	60%	4,224	69%	3,962	70%	199	59%	3,277	68%	2,958	69%	113	64%
Vehicle	1,129	10%	979	10%	76	15%	525	9%	470	8%	49	14%	604	13%	509	12%	27	15%
Other	446	4%	402	4%	24	5%	250	4%	229	4%	17	5%	196	4%	173	4%	7	4%

*The totals for all areas is greater than the addition of non-rural plus rural areas due to fire, deaths, and injuries for which non-rural/rural status is unknown

Appendix D. Average Number of Deaths per Fatal Fire and Average Number of Injuries per Fire with Injuries, 1993-1995

By Property Type	ALL AREAS			NORTHERN AREAS			SOUTHERN AREAS		
	ALL AREAS*	NON-RURAL AREAS	RURAL AREAS	ALL AREAS*	NON-RURAL AREAS	RURAL AREAS	ALL AREAS*	NON-RURAL AREAS	RURAL AREAS
Average Number of Deaths per Fatal Fire									
Total	**1.28**	**1.29**	**1.25**	**1.29**	**1.29**	**1.32**	**1.28**	**1.29**	**1.14**
Outside	1.14	1.19	1.50	1.04	0.91	3.50	1.28	1.50	1.50
Structure	1.29	1.30	1.26	1.32	1.33	1.30	1.25	1.26	1.17
Non-residential Structures	1.40	1.43	1.25	1.48	1.49	1.33	1.28	1.23	1.00
Residential Structures	1.28	1.29	1.26	1.31	1.31	1.30	1.25	1.26	1.17
Vehicle	1.28	1.29	1.21	1.20	1.20	1.22	1.35	1.37	1.00
Other	1.29	1.33	1.00	1.19	1.23	2.50	1.43	1.39	1.00
Average Number of Injuries per Fire with Injuries									
Total	**1.38**	**1.37**	**1.36**	**1.41**	**1.42**	**1.37**	**1.35**	**1.31**	**1.34**
Outside	1.23	1.19	1.18	1.18	1.19	1.17	1.27	1.19	1.27
Structure	1.40	1.40	1.39	1.45	1.45	1.40	1.34	1.33	1.36
Non-residential Structures	1.56	1.57	1.51	1.58	1.59	1.56	1.53	1.53	1.44
Residential Structures	1.37	1.37	1.36	1.42	1.42	1.36	1.31	1.31	1.35
Vehicle	1.34	1.21	1.22	1.20	1.20	1.24	1.47	1.22	1.19
Other	1.37	1.36	1.54	1.41	1.41	1.53	1.33	1.29	1.71

*The totals for all areas is greater than the addition of non-rural plus rural areas due to fire, deaths, and injuries for which non-rural/rural status is unknown

www.ingramcontent.com/pod-product-compliance
Lightning Source LLC
Chambersburg PA
CBHW081615170526
45166CB00009B/2970